AF577330

ISDN AND ITS APPLICATION TO LAN INTERCONNECTION

ISDN AND ITS APPLICATION TO LAN INTERCONNECTION

Derviş Z. Deniz

Eastern Mediterranean University

McGRAW-HILL BOOK COMPANY

London · New York · St Louis · San Francisco · Auckland · Bogotá · Caracas · Lisbon
Madrid · Mexico · Milan · Montreal · New Delhi · Panama · Paris · San Juan · São Paulo
Singapore · Sydney · Tokyo · Toronto

Published by
McGRAW-HILL Book Company Europe
Shoppenhangers Road, Maidenhead, Berkshire, SL6 2QL, England
Telephone 0628 23432
Fax 0628 770224

British Library Cataloguing in Publication Data
Deniz, Derviş Z.
ISDN and Its Application to LAN
Interconnection
I. Title
004.6
ISBN 0-07-707883-7

Library of Congress Cataloging-in-Publication Data
Deniz, Derviş Z. (Dervish Zihni)
ISDN and Its Application to LAN Interconnection / Derviş Z. Deniz
p. cm.
Includes bibliographical references (p. 231) and index.
ISBN 0-07-707883-7
1. Integrated services digital networks. 2. Local area networks
(Computer networks) I. Title
TK5103.7.D46 1994
004.6'6--dc20
93-30050
CIP

12345 CUP 97654

Typeset by Multiplex medway limited, Walderslade, Kent
and printed and bound in Great Britain at the University Press, Cambridge.

To my loving parents

CONTENTS

PREFACE

An evolution is currently taking place in our telecommunication networks: the design, development and implementation of the Integrated Services Digital Network (ISDN). It is by no means a coincidence that the recent advances in electronics, communications and computer technologies have brought the telecommunications and computer worlds ever closer. This is proving to be one of the driving forces in the emergence of the information society. ISDN with its successor, the Broadband-ISDN (B-ISDN), will play a pivotal role in the achievement of this goal.

ISDN will be the cornerstone of a development where computer and telecommunications users in all fields of our daily lives—in education, industry, commerce, government and entertainment—will join the global information technology (IT) village through improved computer and communication standards and services. The concepts of the ISDN are now firmly accepted. The development of the ISDN, the related standards and technology are still part of an ongoing process. The architectural and service aspects of the networks have gained a base from which new applications can be designed.

It is the purpose of this book to look at such an application of ISDNs: at the interconnection of local area networks (LANs) to the ISDN and to each other through the ISDN. LANs are now as ubiquitous as computers, terminals and printers in the office environments. Hence there is a need to study this application for inter-office and homeworker–office communications along with the provision of the ISDN services to the traditional LAN users.

AIMS AND OBJECTIVES

The aim of this book is to provide a look into the data communication aspects of ISDNs and their utilization in conjunction with the existing computer networks in general and to provide an in-depth look into the dynamic channel and bandwidth management in packet-switched communications over circuit-switched lines in particular. The latter is examined in the light of the evolving ISDN standards and some new proposals are made for the signalling protocols necessary to support such an eventuality. Performance of various bandwidth management strategies are evaluated and presented. This book concentrates on the following:

1. *Underlying principles* The book covers a large field in data communications and computer networking in the local and wide areas as well as modelling, simulation and performance engineering. The book tries to unify the common underlying principles in its wide scope. The examples of this are the network interconnection issues in general and in the context of LAN–ISDN interconnection as well as modelling and simulating LAN–ISDN interfaces.
2. *Architectures* The architectural issues involved in the definition of the ISDNs; the network interconnection using the ISO OSI and other networking standards; design of a Dynamic Channel Management Architecture for network layer relays; the building of a connectionless Ethernet–ISDN network layer relay modelling the relay–ISDN interface; and the building of a simulation system are all investigated and reported in the text.
3. *Standards, protocols and services* Wherever possible, the use of global standards and protocols are made. The services available and or obtained are also mentioned.
4. *Modelling, simulation and performance* The area of performance engineering in the design of computer communication and networks is well developed. However, there is little on how to model at reasonable detail the host–network interface as well as to place this interface within the global framework of network interconnection. The problem is that the former necessitates simulation modelling at the micro level, while the latter necessitates modelling at the macro level. An understanding of the issues pertaining to these topics is essential in modelling, simulation and performance engineering.

AUDIENCE

This book is intended for the range of readers who want to develop their understanding of the use of ISDN in real data application environments. This includes students, researchers, designers, developers, network operators working in data communication applications, and other involved in the procurement and management of communications equipment and computer networks.

It is not intended to be a tutorial on every aspect of the ISDN, and as such it deals with the background topics in a concise manner. Readers are given a general introduction to the ISDN, B-ISDN, frame relay, network and LAN interconnection issues. The concept of channel and bandwidth management is introduced and dealt with at greater depth. The issues encountered when using the ISDN as the transport media in a dynamic environment are analysed and presented.

For undergraduate students, this book is more suitable as a second reading material. It is expected that it will be of great value to students on masters' degree courses and research studies, as it covers new material and suggests further research work in many areas.

For instructors and researchers in the industry, it will be a useful reference for the coverage of a broad range of topics, from ISDN to network interconnection to dynamic channel and bandwidth management to modelling and simulation.

Designers and developers will find the book useful for its concise background reading on ISDNs and network interconnection and its detailed approach to dynamic channel and bandwidth management, design of network layer relays, computer and LAN–ISDN interfacing.

ORGANIZATION OF THE BOOK

The book can be considered in four broad sections:

- *ISDN* This section introduces the ISDN, which is the new global digital telecommunications facility, and deals with the pertinent issues in data communications using ISDN.
- *Network interconnection* The general network interconnection issues encountered in the computer communications world are considered.
- *Channel management, architectures and protocols* This section deals with the channel management issues encountered when a multi-channel facility is available at the network access point. It concentrates on the case of the packet-switched data communications over circuit-switched links with a particular reference to the circuit-switched ISDNs and their use in LAN–LAN interconnection.
- *Dynamic bandwidth management policy* Queue- and rate-based dynamic bandwidth management policies are studied and compared. Possible applications are investigated.

The organization of individual chapters is presented in more detail at the end of the introduction.

ACKNOWLEDGMENTS

There are many people whom I wish to thank for all their help and support. Most of the results used in this book have stemmed from my previous research undertaken at the University College London (UCL). Therefore, first, thanks must first go to Professor Peter T. Kirstein, Head of the Department of Computer Science, for his continuous support and permission to conduct further research at UCL, and to Graham J. Knight for his valuable discussions and suggestions during the original phase of the work.

I also wish to thank many other staff members and colleagues in the Department of Computer Science at UCL (UCL-CS), for their valuable discussions, encouragement and friendship. Thanks are due especially to Dr Soren A. Sørensen for his most helpful suggestions and encouragement in the simulation phase of my research; and also to Jon Crowcroft of UCL-CS and Dr Valerie Isham of the Statistics Department at UCL for valuable discussions.

I duly acknowledge that the core of this work has been made possible only by the support of the Association of Commonwealth Universities, the British Council and the Department of Computer Science, UCL. Thanks are also due to the Eastern Mediterranean University (EMU) in the Turkish Republic of Northern Cyprus for the arrangement of my research leave. I also wish to thank Mehmet Gencer, Mustafa Buzun and Attila Alişkan at EMU for their help in the production of the index.

It would have been impossible to bring this work to light without the unending support of McGraw-Hill's staff at their Maidenhead offices. Therefore, I would like to thank all the coordinating staff and editors who have made a marvellous job out of the original manuscript.

CHAPTER

ONE

INTRODUCTION

The integrated services digital network (ISDN) is one of the hottest buzzwords today, just as the micro-chip was about fifteen years ago. A lot of development has taken place since then, in both the computing and the communications worlds. Imagine a computer sitting on your desk with which you could ring anywhere in the world and carry on a telephone conversation while at the same time accessing a remote database from within another window on your screen and using the same wall socket: your introduction to the world of ISDN. Imagine yourself then making a call to your local estate agent and scanning through their home catalogues on your screen while discussing the details with them over the phone connected to the same twisted pair cable. Imagine yourself in the office conducting a multi-party, multi-media conferencing across continents. Well, all of this technology is currently available, thanks to ISDN and the developments in data and telecommunications techniques and standards.

Local area computer networks (LANs) have become as ubiquitous in today's offices as personal computers are at home. Formerly, the interconnection of LANs to each other left a lot to be desired in that they mostly used the old analogue telephone lines and modems operating at the now abysmally low transmission rates of 9600 bits per second bps. Packet-switched data networks (PSDNs) have alleviated some of these problems, but the technology and applications in the LAN world have also outgrown their original bounds. What is currently needed is a method of transparent interconnection of LANs and of porting the applications developed for the relatively fast and error-free environments found on the local area networks across long distances over the ISDN. These services should be available to home workers, office interconnects and global LAN interconnection. Dynamically varying network topologies and virtual networks are now possible, thanks again to the ISDN. Faster speeds, a very good bit error rate (BER), end-to-end digital connectivity, support of multiple services and types of services, standardized access methods and standardized hardware and software are now a reality. For voice, data and video services supported concurrently through one global network and ubiquitous access all around the globe, the ISDN is the way forward; and, what is more, the technology is available today.

This book is an introduction to the ISDN and its application to the interconnection of local area networks. It aims to investigate the application of this new technology to computer communications. Most of the information contained in this book is derived from direct research and as such consists of some material that is new and is still evolving. Standardization is being proposed in several of the areas discussed. Furthermore, new material and methods are introduced with reference to the existing and developing standards in both the computer and the telecommunications worlds. The book aims to cover a gap in the ISDN literature by showing the methods used in the computer applications over ISDN and the recent developments in the technology. Communications architectures and protocols are also presented. One of the main problems encountered in the narrowband ISDN when used to interconnect remote LANs has been the dynamic management of the multiple channels at the user–network interface. This issue is taken up and analysed in depth with reference to actual implementations and dynamic management of the communications bandwidth in the latter parts of the book. Although the latest developments in the ISDN world are reported, the book concentrates mainly on the applications of the circuit-switched services over *narrowband* ISDN.[1]

1.1 EMERGENCE OF ISDN

The invention of the transistor and the subsequent development of microchip technology has resulted in a variety of useful applications in almost every aspect of life; from home to office, from industry to commerce, from government to health care to education. This is currently progressing in two main fronts: the computer and communications technologies. When intelligent functionalities were increased in the old telecommunications networks, it came in the form of the stored program control (SPC). This was the use of some computing function running special software to control the exchanges in the network. A multitude of services then became available—call logging, itemized bills, transfer of calls, conferencing, ring-back-when-free, etc. On the other hand, it became possible for computers with simple interface cards and modems to access the telecommunications services through their communications ports and software. Slowly, a bit of the functionality of each of these technologies became a part of the other. Today they are inseparable in many respects. Coupled with this background development are the user requirements of accessibility, flexibility, speed, cost effectiveness and new services on the one hand, and the desire of the service providers to increase the return from their invested capital, to use the economy of scales, to increase serviceability and to ease the management problem on the other. Hence the technological and market forces have required the integration of existing networks and services already provided on separate networks. This has led to the concept of an integrated services digital network, a network that would provide a solution to most of the problems in the communications world.

[1] In the rest of this book, the term 'ISDN' will be used to mean the *narrowband* ISDN (N-ISDN), and B-ISDN will be used specifically to mean *broadband* ISDN (see Chapter 2 for details). Note that, while the N-ISDN access is based on the time division multiplex (TDM) technology, the B-ISDN is based on the asynchronous transfer mode (ATM). The N-ISDN has two access structures: basic-rate access (BRA or BRISDN) and primary-rate access (PRA or PRISDN). The BRA and PRA interfaces are composed of 2B+D and 30B+D (in Europe) channel structures, respectively. The B-channel has a transmission capacity of 64 kbps. The D-channel has 16 kbps and 64 kbps transmission capacities for the BRA and the PRA interfaces, respectively. The PRA additionally has higher capacity channel structures (see Chapter 2).

There are two main aspects of ISDN: the network evolution, and the services provided. As far as the users are concerned, ISDN will provide all the necessary communications services or access to the services provided by other networks in a transparent fashion. As far as the service providers are concerned, the services may be provided by one homogeneous or several heterogeneous interworking networks. This does not change the view of the ISDN from the outside: that of a uniform service provider through standard interfaces.

The integrated services digital network (ISDN) is evolving from the integrated digital network (IDN) concept and is taking shape worldwide. The IDN provides the integration of the switching and transmission facilities and extends it to the subscriber loop by digitization in the network. It also provides for the common channel signalling which is based on the transmission of the control and signalling messages on a packet-switching network designed for this purpose and is part of the public switched telecommunications network (PSTN). One of the fundamental concepts in the creation of ISDN is the provision of a multitude of switched and non-switched services to the users within the circuit, packet of frame modes of access, through the use of a small set of standard user–network interfaces (UNIs). Apart from fast switching, which is possible because of the end-to-end digital connectivity, the ISDN utilizes common channel signalling (CCS), allowing the selection of different services through the use of a standard signalling protocol at the UNI. Furthermore, the ISDN provides multiple channels to the user at the UNI. The main multiplexing technology used in the digital telephony world is time division multiplexing (TDM). Multiple channels at the ISDN user–network interface are formed by assigning one or more time slots (TSs) within a TDM frame to a channel. The TDM technology lends itself easily to circuit switching (CS). Hence, circuit switching is *inherent* to ISDN, and it is one of the earlier services available in ISDNs.

More recent technological developments taking place in the ISDN world are the frame mode services, based on the more efficient use of the ISDN technology, and the broadband ISDN services, based on the asynchronous transfer mode (ATM) and the fibre optic technology. These will bring a drastic improvement in the way most current services are provided. For example, the frame mode services are proposed as the main method of LAN–LAN interconnection in the very near future. The ATM technique is proposed as the main technology for multi-megabit services including high-definition TV and video conferencing. However, in this book we shall be concentrating on the existing narrowband ISDN and the circuit-switched bearer service.

Some of the factors affecting the development of the ISDNs are:

- *Basic technology* Among developments in the sophistication of electronic components, the current very large-scale integrated circuit (VLSI) chip technology allows the implementation of many more functions in hardware and firmware. For example, silicon chips are available accommodating the high-level data link control (HDLC) protocol for layer 2 of the OSI reference model (CCITT 1988a; ISO 1984). Developments in the fibre optic technology is another factor.
- *Increased use of computers* Computers are nowadays used in almost every walk of life, including government, commerce, banking, education, research, industry, and tourism and leisure. The development of sophisticated applications requiring more and more hardware capacity and communications bandwidth means that the computer and telecommunications technologies have to deliver. The availability of on-line information services and large databases, and the demand for electronic shopping and banking

facilities, mean that more and more data networks will be built. Although currently the data services amount to only a very small percentage of the whole telecommunications sector, this is bound to change in the near future.

- *Increased use of communications services* The whole public and commercial life of individual countries and companies nowadays depends more than ever on the use of communications facilities. This provides instant access to information on a national and global basis. The whole trade and commerce sector is now completely dependent on the availability of telecommunications and computing services. For example, one cannot conceive of a just-in-time type stock provision without telecommunications and computing facilities.
- *More time and money to spare* With current developments in the economic and social spheres, most people will have more time and money to spend on technology and leisure. They will want more TV channels and easier access to electronic services and teleconferencing facilities.
- *Working from home* With the increase in computing and communications facilities, 'home commuters' are growing in number. This especially suits people who prefer to work at their home rather than the office environment or who have a good reason not to travel. It also suits some businesses, allowing them to cut down on office space, heating costs, etc.
- *Information technology* The current state of information technology has reached such a level that the storage of information is no longer the limiting factor; rather, it is the access and manipulation of the information. Hence faster and better communications facilities are needed.
- *Computer-aided production and support* Most industrial output is nowadays controlled by communicating computers. The move is to lock the parts manufacturers, stock suppliers, engineering design, manufacturing and servicing centres, as well as the distributors, into one work environment based on suitable standards, software, services and a global computer and communications network. An example of this is the Computer-Aided Logistics Support (CALS) programme of the US Department of Defense (LMI 1989, 1990).

1.2 THE DATA COMMUNICATIONS REVOLUTION

A revolution has been taking place since the early 1980s whereby the individual computer with its own domain of information and work is no longer a preferred solution to computing. Today, computers have become much more friendly and accessible. They have become tools for productivity rather than simply pieces of scientific equipment. Distributed computing based on workstations, personal computers and their networks has become the norm. This has resulted in the proliferation of the local area computer networks (LANs). Next has come the departmental LAN, connecting whole departments, and the company LAN, formed by the interconnection of the departmental LANs throughout a company. In parallel, a more widespread revolution has taken place; that of the internet. The interconnection of company or institution LANs to each other via a special wide-area computer network using special communication protocols has meant that a concatenated network of networks can be formed. This is called an *internet* if they all support a common internet protocol at the network layer of the Open Systems Interconnection (OSI) reference model.

Another development in the public domain has been the adoption of network access standards by world standards organizations for the use of national and international packet-switched public data networks (PSPDN). An example to this is the CCITT Recommendations on the X-series of protocols (e.g. X.25 and X.75). Parallel to this, other standards organizations, such as the International Standards Organization (ISO) and the Institute of Electrical and Electronic Engineers (IEEE) in the United States, the British Standards Institute (BSI) in the UK and AFNOR in Germany, have been active in the definition and acceptance of other data communications and networking standards. One of the most important sets of standards are the ISO's Open Systems Interconnection (OSI) standards on computer communications (Day and Zimmerman 1983). OSI standards cover only the data applications. However, work in the telecommunications world is converging to that of the computer world. Therefore, standards covering the overlapping area need to be defined. Indeed, today more than ever before, there is a need to provide standards for multi-service and multi-media networking.

Today's networking covers a wide spectrum. It involves computer-to-computer as well as computer-to-other-digital-device communications like disk storage devices, tape drives, printers and even industrial machinery. To cope with specific applications, Technical and Office Protocol standards (TOP) (see Boeing 1988) and Manufacturing Automation Protocols (MAPs) (General Motors 1988) have been developed in some sectors of the industry.

Computer communication is necessitated by various needs. Some of these are:

- *To share data and information* Inaccessible data is not useful. The ability to share information and exchange data is vital in today's work environment.
- *To share resources* Expensive and vital resources can be shared between users and computing machinery. Examples are the expensive supercomputers, laser input or output devices, line printers, tapes and disks.
- *To increase reliability and extend services* A distributed system can be made to have increased reliability over a single resource as a result of the replication and extended services.
- *To control remote devices* Sometimes devices that need to be controlled by computers can be situated at long distances from the control centre. In this case a computer network may be installed with the appropriate hardware and software. Data collection and transmission is also required.

Many computer applications are now available using computer communications and networks. These include electronic mail, remote login, file access, file transfer, remote windowing, remote database access, computer conferencing and computer-aided telephony.

1.3 GLOBAL IT VILLAGE

From the preceding sections, it is clear that the current trend in the computer communications revolution is towards global access to information and exchange of data using existing and future applications. This will soon create a 'global IT village', where the wonders of information technology (IT) will be exploited to its fullest. We will be able to use computers to access huge data banks, transfer information through networks at the speed of light, and request multi-media conferencing including voice, data and video components. We will have access to ever more information through document delivery services and will be able to do our banking, shopping, education and working from home. Access to these services from virtually anywhere will be possible using mobile voice and data as well as mobile LAN/MAN services. Storage, retrieval and processing of data, voice and video messages located on special servers will be possible. The world is going to shrink in size still further in terms of global communications and information dissemination.

All this and more are heralded by the latest developments from the technological frontiers. However, first we must come down to search and investigate how we could utilize the flexibility and opportunities provided by the emerging ISDN. One of the requirements is to port existing services on to the ISDN. In the case of computer applications, an urgent need is to port a multitude of services (like LAN interconnection, remote windowing and networked file systems) that are by now so accustomed (and are taken so much for granted) by so many computer and LAN users to the ISDN.

1.4 INTERNETWORKING IN THE ISDN ERA

1.4.1 The background

The first computer systems consisted of a mainframe forming the central hub of a star network serving user terminals and other peripheral equipment. The communications between a central computer and its peripherals formed the first data communications as such. These islands of computing systems were soon found to be wanting in many respects. The need for computer communication arose mainly from the need to share information and resources. This led to the formation of local computer networks. However, the need for inter-site, inter-city, inter-country and global data communications has resulted in the establishment of private, national and international computer networks. Examples of these are the networks of individual banks or companies (e.g. IBM's VNET and DEC's Easynet—see Quarterman and Hoskins 1986), national research and educational networks (e.g. ARPANET in the United States (Leiner *et al.* 1985), JANET in the United Kingdom (Kirstein *et al.* 1989), European Academic Research Network (EARN) and RARE (Quarterman and Hoskins 1986)).

As computers entered every aspect of human, economic, production and administrative activity, so too did the industrial, commercial, financial, educational, governmental, health and even recreational applications of computing become widespread. As the number of computer networks at the local or wider areas increased, the need for interconnectivity and interworking between different networks is becoming more acute.

For data communications between computers, a set of procedures, rules and conventions defining different activities needs to be established. These are called the communication *protocols*. Most networks are designed as a series of *layers* or *levels*, each one built upon its predecessor. This provides structured design with reduced complexity. A set of layers and protocols is called the network *architecture* (Tanenbaum 1989). The International Standards Organization has been working towards the establishment of world-wide standards collectively known as the Open Systems Interconnection (OSI) standards, based on the principle of layering. This principle helps the structured design of computer networks. The OSI reference model (OSI-RM) (CCITT 1988a; ISO 1984), defines seven layers and is used as a guide for all network architectures conforming to the open system principles (see Fig. 1.1). Open systems are systems that are open for communication with other systems conforming to the same principles.

In the OSI-RM, each system is decomposed functionally into a set of subsystems and is represented pictorially in a vertical sequence. Vertically adjacent subsystems communicate through their common interfaces, while *peer* subsystems collectively form a layer in the architecture. Each layer provides a set of well defined services to the layer above, by adding its own functions to the services provided by the layer below (Day and Zimmerman 1983; CCITT 1984). The layers of the model are partitioned as follows:

- *1: Physical layer* Achieves the transmission of raw data bits over a communication channel (medium).
- *2: Data link layer* Converts the raw transmission facility into a line that appears free of transmission errors to the network layer. This is done by grouping level 1 bits into data frames and delimiting them. This layer may also include access control to the medium, error detection and correction.
- *3: Network layer* Performs the routing and switching of data between any two systems across multiple data links and subnets.
- *4: Transport layer* Operates on an end-to-end basis achieving the necessary quality of service for the exchange of data between two end systems. May include end-to-end error recovery and flow control.
- *5: Session layer* Allows users on different machines to establish sessions between them, and hence establishes and manages communication dialogue between processes.
- *6: Presentation layer* Manages and transforms the syntax of structured data being exchanged. Is also concerned with the semantics of the information transmitted.
- *7: Application layer* Deals with the information exchange between end-system application processes and defines the messages that may be exchanged.

The above layering was created according to the original design principles used in the construction of the ISO model. According to this (see Day and Zimmerman 1983):

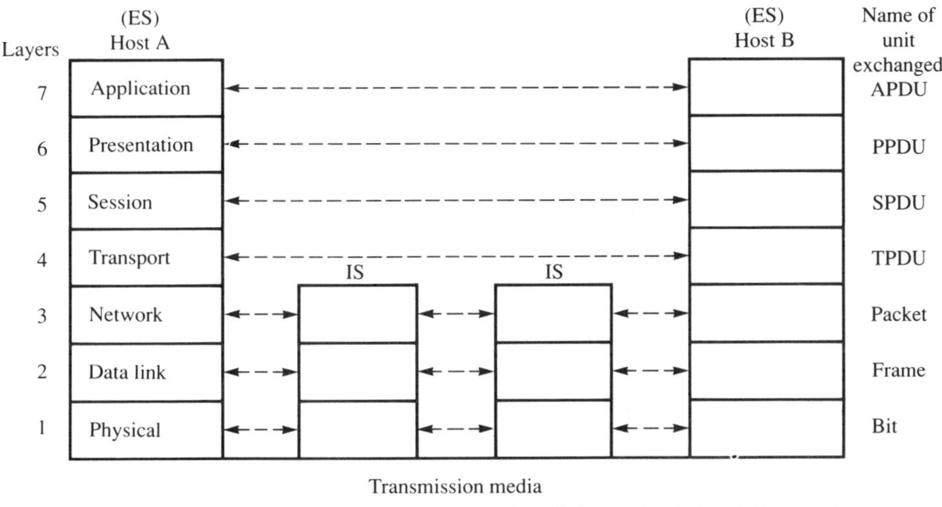

Figure 1.1 The OSI reference model

1. Different levels of abstraction are placed in separate layers.
2. Similar functions are grouped together within a single layer with each layer performing a well defined function.
3. The function of each layer is chosen so as to be amenable to the definition of a standard protocol.
4. Minimization of information flow across interfaces is a primary goal in drawing the layer boundaries.

Although the majority of network architectures widely in use are based on the principles of layering, most do not fit the OSI model exactly in their allocation of layers and protocols used. Examples of these are the IBM's SNA (Meijer 1987), DECnet and DARPA Internet (Quarterman and Hoskins 1986), to name but a few. Conversely, some new network architectures, such as the MAP (General Motors 1988; O'Prey 1986) and TOP (Boeing 1988), have adopted the OSI-RM for their architecture and hence form 'open' networks. Open networks use internationally standardized procedures for communications rather than local or proprietary ones.

1.4.2 Technological base for internetworking

In computer communication three basic types of switching are used: packet, circuit or message switching. Most data networking is packet-based since, whatever the underlying switching or transmission mode used, some form of data framing is used. A *data frame* may be considered to be the smallest unit by which data transfer between networked elements is achieved, since data bits are grouped into frames before transmission. The packet size may be greater or smaller than the frame size. Packetization of data into units makes the fragmentation, transmission and reconstitution of the original data easier in that each unit may be accounted for by the communication protocol used. Hence, error detection and correction can be handled more easily. Also, flow and congestion control can be more manageable.

In packet-switched (PS) networking, hosts or network relays have single- or multiple-access ports connected to packet switches in the main subnet (or WAN). Transmission links between the packet switches and the access lines to the packet switches are usually permanently connected via leased or private lines.

Two modes of operation are possible in PS data communications: connection oriented (CO) and connectionless (CL). In the CO mode, first a connection termed a virtual circuit (VC) is set up across the packet-switching network to the destination. All the intermediate switching nodes must reserve resources for the new VC. Before the establishment of a VC can be achieved, a data link layer (DLL) connection must be established. This is usually done at the start-up time. Network layer (NL) packets are then transferred in DL frames.

In the CL mode of operation, no previous connection set-up is needed and network layer packets are individually routed to any one of the transmission lines over which a data link connection exists to the destination. Transmission of network packets next hop along the route is achieved by transporting them in DLL frames.

Table 1.1 Access methods

Transmission	Access	Mode
Analogue	Acoustic coupler	Switched
	Modem	Switched/leased line
Digital	Direct	Switched/leased line
	Integrated access	Switched/permanent/semi-permanent

In circuit-switched (CS) networking, network access is achieved only after a circuit has been set up to a destination, through the intermediate network. If a packet service is required, then, once the physical layer connection is achieved, a data link connection needs to be established between the two ends. The intermediate network has no knowledge of the framing or packetization, acting solely as a 'bit pipe'. Network packets can then be forwarded between the two communicating ends. In the CS mode of networking, the delays in the establishment of physical circuits and DLL links can be time-consuming, preventing fast, on-demand set-up and removal of connections. Tables 1.1 and 1.2 show the access modes in the analogue and digital networks and the differences in the packet- and circuit-switched networking, respectively.

Table 1.2 Comparison of packet-switching and circuit-switching networking

PS	CS
No 'circuit' needs to be set up (CL) VC must be set up (CO)	Long circuit set-up delays
Point-to-point, multicast, broadcast	Point-to-point
Longer queuing delays may be encountered and packet-switching delays at every switching node *en route* suffered	Once the circuit is set-up, transmission and propagation delays are the most important delays encountered
Multiplexing is inherent	Multiplexing is more convoluted

In the ISDN era, fast circuit switching will be possible with end-to-end circuit set-up times of less than 1 s within a country and using terrestrial links. (Longer distances may take up to 4.5 s on average.) Hence, with the ISDN technology, on-demand circuit establishment and removal will be possible. This will lead to a novel way of operation using the circuit mode bearer services where the number of channels established to a destination can be varied dynamically as a function of traffic. If the traffic to a destination increases, causing the number of packets in a transmission queue to build up, additional channels can be established to the same destination in order to reduce packet delays. Similarly, if the traffic to a destination decreases, one or more channels to that destination can be disconnected, minimizing costs.

Table 1.3 Comparison of existing networks and ISDN

Features	Existing networks	ISDN
Access	Separate network access to telephone, CSPDN, PSPDN, telex and teletext networks	Integrated digital access to bearer and teleservices
Signalling	Different in-band signalling for each network	Out-band common channel signalling for access to all ISDN services
Channels	CSPDN: separate 'lines' needed PSPDN: up to 1024 VCs on each 'line' is available	Multiple channels, each of which can be used for access different services and destinations concurrently
User protocols	CSPDN: usually X.21 at the PL PSPDN: X.25, LAP-B	CS-BS: ISDN protocol I.431 (PL); PS-BS: X.25, LAP-B; APM-BS: frame relay/switching; (LAP-D)
Switching speed(s)	PSTN: up to 30s CSPDN: < 1 s PSPDN: fast VC set-up times	< ls (typical 500 ms) end–end (terrestrial short distances); worst-case terrestrial: 4.5 s (mean)
Transmission speed(s)	Switched: up to 64 kbps Non-switched: 64 kbps or higher	Switched or non-switched: 64 kbps or higher
Availability	Separate subscription to services is needed, necessitating multiple sockets	Can use existing telephone network wiring giving widest coverage; single access point to all services

Another incentive for disconnecting unused connections is the tariff. Current pricing policy of postal, telephone and telegraph companies (PTTs) mean that ISDN CS calls cost the same as telephone calls (or a simple multiple). This means that, after the basic connection charges (if any), the charging for usage will be based on distance, duration and the time of day. (See also Section 2.10.) By contrast, packet-switched networks charge on the volume of data sent in each direction, plus connect time. Therefore, ISDN CS service will cost money even if the line is not fully utilized, as the capacity must be allocated to circuits even if they do not carry traffic. The issue of dynamic channel management as introduced here will be the topic of investigation both in Section 1.5 and in Chapters 4 and 5.

1.4.3 Using the ISDN

ISDN services At the user–network interface (UNI), the ISDN (CCITT 1988a) provides access to both *bearer services* and *teleservices*. Bearer services provide only the basic communications services (bit transmission), whereas the teleservices provide access to value added services and other network services like telephony, e-mail, voice-mail and telex. Any of these services can be requested through the use of ISDN signalling protocols on the D-channel. The bearer services further subdivide into circuit, packet and frame mode bearer services, providing circuit-, packet- and frame-based channels to the user, respectively.

Access types Two types of access are defined in the ISDN: basic-rate access (BRA) and primary-rate access (PRA). The basic-rate access provides two user channels at 64 kbps and a signalling channel at 16 kbps on a twisted pair to the user. The primary-rate access provides 23 (in the United States and Japan) or 30 (in Europe) user (B-) channels and 1 signalling (D-) channel, each with a transmission rate of 64 kbps. Higher-rate channels, e.g. H0 channel, are also available (see Chapter 2). A coaxial cable is used for each direction of transmission.

Impact of ISDN services Two types of interworking are envisaged between data and telecommunications equipment and the ISDN:

1. The use of ISDN as a transparent network to interconnect existing data networks by using only the bearer services of ISDN: this leads to the data-communications-oriented (DCO) interworking.
2. The use of ISDN teleservices and innate applications: this results in the telecommunications-oriented (TCO) interworking.

The differences between these two types of interworking result in different complexity requirements from the network attachment units. In DCO interworking, only the bearer service access and control parts of signalling protocols are needed. On the user channels, two options exist: the use of ISDN data mode access protocols (currently only X.25 is supported), or of user-defined protocol stacks over CCITT defined physical layer protocols. In the TCO interworking, additional signalling protocol features are needed for access and control of teleservice sessions. On the user channels, CCITT-defined protocols must be used. Both of these ISDN interworking types are dealt with in more detail in the next chapter.

Data communications over ISDN The CCITT Recommendations (Recs.) of the I-series describe four types of data services: circuit, packet and frame modes and the support of data terminal equipment (DTE). The CS data service provides only a ‘bit pipe’ down which data bytes can be sent, and it imposes no user protocol restrictions apart from the ISDN Physical Layer Protocol (CCITT 1988b, 1988c). The packet mode data service specifies three types of services: user signalling, connectionless and virtual circuit services. The frame mode service (also called the ‘additional packet mode bearer service’—APMBS) supports frame relaying, frame switching and X.25 protocol operation. Support of DTEs provide access to the X- and V-series DTEs and packet mode DTEs (e.g. X.25 and ISDN-compatible terminals). Packet mode DTEs can provide access to PSPDN services and can utilize the ISDN VC service. While the public switched packet data network (PSPDN) access defines the lower three layers of protocols, the APMBS defines only the lower two layers of protocols when frame mode services are used and offers the lower three layers when X.25-based services are used.

1.4.4 The LAN–ISDN interconnection[2]

Users connected to the existing data communication networks may want to interconnect their local networks or individual workstations through the ISDN, or may want to interwork with the ISDN. The first case implies that the users want to use only the bearer service of the ISDN as a 'raw carrier'; the second case implies that the users want to access the teleservices as well as the bearer services. These two options present different problems for interconnection and necessitate the DCO and TCO interworking, respectively.

Most of today's local data networks can be modelled as multiple hosts, workstations, personal computers and terminals interconnected through a local area network. The need to interconnect these LANs with other LANs at remote sites has already been discussed. In the ISDN era, DCO or TCO interworking through public or private ISDNs is possible.

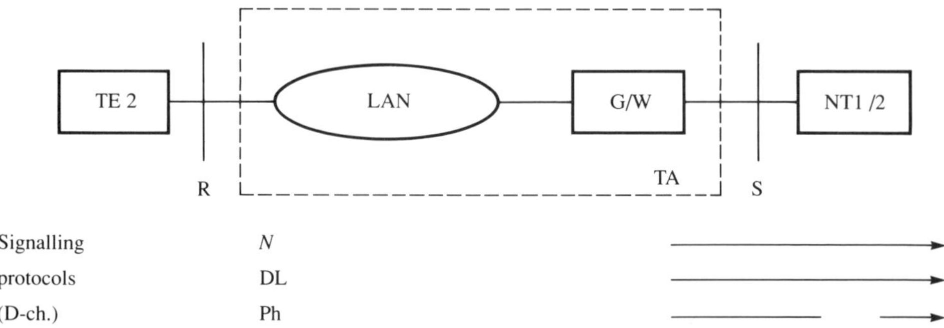

Figure 1.2 DCO interworking : LAN and gateway (G/W) as a TA

A private ISDN network may be a local integrated services private branch exchange (ISPBX) or multiple ISPBXs interconnected by leased lines to which multiple LANs can be attached. Hence, the two scenarios of interworking needs to be investigated. Figure 1.2 shows the DCO interworking scenario, while Fig. 1.3 shows the TCO interworking scenario for LAN–ISDN interconnection. The main difference between the two is in the scope of the signalling protocols. In the DCO interworking, the ISDN is made transparent to the LAN hosts. It is this mode that applies to the LAN–ISDN–LAN working since traditional LANs are data-communications-oriented. In the TCO interworking, the signalling protocols need to be terminated in the LAN hosts, making the LAN transparent to ISDN-compatible terminals and ISDN services.

[2] In this book a distinction is made between the LAN–ISDN and LAN–ISDN–LAN interconnections. In the broadest sense, the former can be assumed to support integrated services facilities of the ISDN. However, since most existing LANs support only data services, and since this book is concerned mainly with data communications, it will be assumed to refer to data applications only unless indicated otherwise. In the LAN–ISDN interconnection it is further assumed that remote ISDN workstations can access the LAN hosts and services transparently. In contrast, the latter implies that the ISDN is transparent to the LAN applications and supports only data communications. This means that the two end LANs are more tightly coupled, and the hosts and workstations attached to the ISDN may or may not be able to access the LAN services. (See Chapter 3 for further discussions on this topic.)

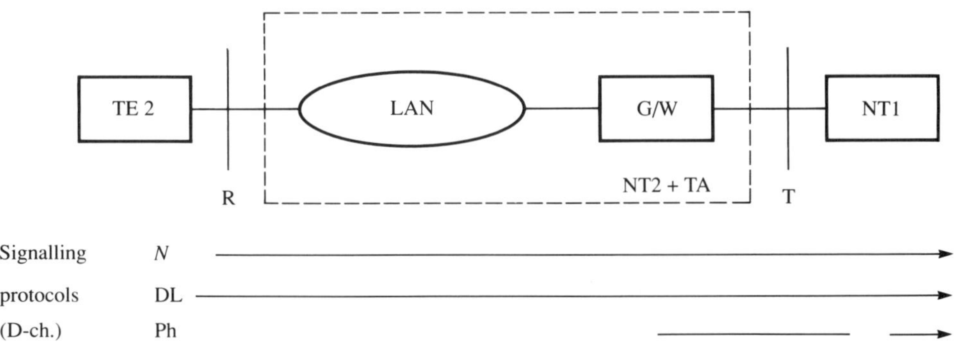

Figure 1.3 TCO interworking : LAN and gateway (G/W) as a NT2 + TA

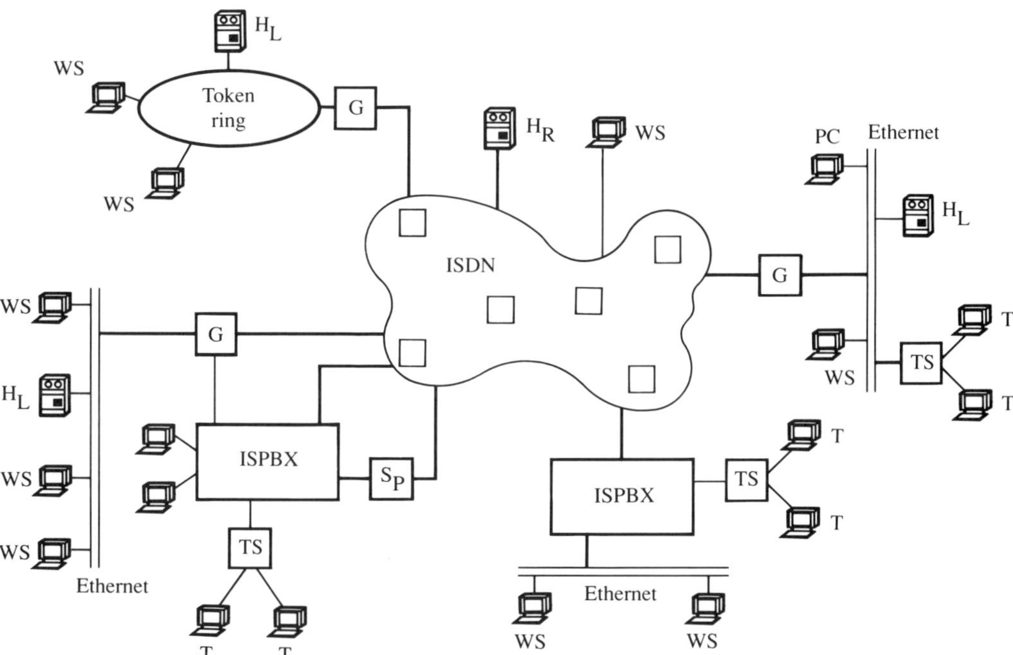

S_P: primary-rate server; T: terminal; TS: terminal server; WS: workstation
PC: personal computer; H_L: local host; H_R: remote host

Figure 1.4 An interworking reference configuration

These issues are further discussed in Chapters 2 and 3. Figure 1.4 shows a reference configuration for LAN–ISDN and LAN–ISPBX–ISDN as well as a workstation–ISDN interconnection.

1.4.5 Future trends in network services and networking

The common underlying trend in the network services provision, whether at the local, metropolitan or wide area, the public or private domain, is the integration of services. This implies the provision of voice, data and video services within an all-encompassing network resulting in the integration of services as seen by users at the access node. In earlier stages, the provision of actual services may be done by different networks, but the user need not be aware of this. ISDN is following this route as it is an evolutionary rather than a revolutionary network. The concept of services integration is driven by two main objectives:

1. Provision of a simplified access to a multitude of services through a single interface—integration of services from the user perspective.
2. Provision of network integration such that implementation and operational costs can be reduced owing to economies of scale. For example, it is expected that the cost of provision of data services in ISDN will be met by revenues from the telecommunications users, making the provision of data services within the ISDN cost-effective for PTTs. This is the integration of services from the perspective of service providers.

The problem with voice, video and data integration is the vastly varying requirements of these services. Voice requires bounded delay with low coefficient of variation (low jitter) and is better served by a synchronous communications channel (Hilal and Liu 1984). Some loss is also tolerated. Data requires error-free delivery, but can tolerate delay. Video requires much greater bandwidth than either voice or data; it also requires bounded delay, while some loss is tolerated. Compression techniques for both voice and video are available, but this necessitates better bit error rate (BER) communications, since any loss of information becomes more prominent. Again, video is best served by a synchronous communications channel.

A parallel development is the arrival of multi-media services. As opposed to the integrated services, these services incorporate multiple forms of information representation. Examples of such services are the voice-annotated text service and text with bit map diagrams/pictures. Many more will be available in the future. Most of the documents exchanged by these services require huge bandwidth capacities for transfer.

Although no international standards exist for services integration at the local and metropolitan area networks, work is carried out by various national and regional institutions for services integration. An example of this is the integrated voice and data (IVD) architecture study by the IEEE (1988a). Services integration can be further studied at the local, metropolitan and wide area networking levels, as it affects all three of them.

Local area networking The controversy of LAN versus private branch exchange (PABX or PBX) in the provision of local area networking has been going on for years. It seems that both are going through an evolution and will coexist together, filling similar roles and interworking. Integrated-services LANs (ISLANs or ISLNs) (Deniz 1985b), as well as integrated-services PABXs (ISPBXs), have so far concentrated on the integration of voice and data. Traditionally, LANs have been better at dealing with data, while the PABXs are better at dealing with voice.

Various proposals exist for ISLANs. These either involve modifications to the existing media access protocols, so as to provide the service requirements of different services, or propose new technologies. Examples of the first type are the Orwell ring (Adams and Falconer 1984; Falconer and Adams 1985) (based on slotted ring), the modified Ethernet (Gonsalves 1983; Nutt and Bayer 1982; Tobagi and Cawley 1982) or token rings (Ibe and Gibson 1986; Wong and Gopal 1984), while FDDI-II (Ross *et al.* 1990) is of the latter type and is based on fibre optics.

ISPBXs are already available and implement ISDN access protocols for both the user and network interfaces for compatibility. In the United Kingdom a similar protocol for ISPBX–ISDN access exists and is called the DASS-2 protocol (British Telecom 1985). However, for the ISPBX–ISPBX interconnection there are as yet no international protocols. In the United Kingdom a protocol called the digital private network signalling system (DPNSS) (British Telecom 1989) exists and is based on the common channel signalling (CCS) principles. In Europe there are moves by France and Germany to produce an alternative protocol. The interconnection of data processing equipment and PABX is covered by a technical report (TR-24) produced by the European Computer Manufacturers Association (ECMA) (Leiner *et al.* 1985).

LAN (or ISLAN) and PABX (or ISPBX) interconnection is another issue where no international standards exist. A study by IEEE (1988a) on the LAN–ISDN interconnection proposes the use of frame relay.

Metropolitan area networking (MAN) Integrated services MAN is now becoming feasible. An example is the distributed queue dual bus (DQDB), which is a draft standard in the form of IEEE 802.6 (IEEE 1988c). These will provide access to synchronous and asynchronous services and hence provide voice, video and data integration at the metropolitan area.

Wide area networking (WAN) ISDN is a new concept that is providing a useful framework for the development of future telecommunications networks and services. The natural extension to ISDN is the broadband ISDN (B-ISDN). CCITT Rec. I.121 (CCITT 1988a) specifies the asynchronous transfer mode (ATM) as the technology on which the B-ISDN will be based.

ATM technique will provide synchronous as well as asynchronous services. Access to ATM networks will be through fibre optic links, and so far 150 Mbps and 600 Mbps access rates have been defined (CCITT 1988a).

1.4.6 New opportunities

The ISDN standardization has come of age; the first CCITT Recommendations relating to ISDN, the Recommendations of the I-series, were released in 1984 (Red Books) as a result of the first study period 1980–4 by the Study Committee XVIII (CCITT 1984). The second study period of 1984–8 culminated in the updating and extension of I-series Recommendations (Blue Books) (CCITT 1988a). The Red Book seems to have approached the ISDN issue from the telephony side of services and hence only specifies access to the PSPDNs using X.25 within X.31 scenarios. The Blue Book seems has taken a closer look at the data services and hence incorporates sections on new packet mode services as well as indicating the selection of ATM technique for B-ISDN. Despite all this, the network implementations and development of applications have been slow to follow. Currently some commercial services are available in a few industrialized countries (United States, United Kingdom, France, Germany, Japan, etc.), and several others are in the pilot phase of their implementations (Italy, Spain, etc.). A

Memorandum of Understanding signed in 1989 by the EC members, Turkey and Yugoslavia (CN & IS 1991) is expected to achieve a European-wide ISDN service by the end of 1993.

The CCITT released its Interim Recommendations on ISDN in 1990. These include work that was carried out during 1988–90. Among the topics studied are the ISDN frame mode bearer services (FMBS) (I.2xy), the ISDN Protocol Reference Model (I.320), ISDN network architecture (I.324), network performance objectives for packet mode communication in ISDN (I.35p), FMBS interworking (I.5xz), and congestion management for frame-relaying bearer service (I.3xx). More work items were finalized in the CCITT Plenary Assembly that took place in Helsinki (1–12 March 1993).

One of the purposes of this book is to address the important issue of dynamic channel management. The approach adopted here recognizes the need for transferring current data applications to ISDN. Furthermore, new features of ISDN are attractive in that they promise much more flexible usage of resources. There is a gap in the existing networking and the defined features of ISDN; hence I identify the following needs in data applications of circuit-switched ISDN:

- Managing the multiple channel resource effectively
- Converting from traditional data in-band signalling to ISDN out-band signalling
- Defining a software interface between data applications and the ISDN signalling protocols.

1.5 DYNAMIC CHANNEL MANAGEMENT

1.5.1 Definitions

Resource management is a well-known concept in data communications and computer networks. It enables users or applications to manage the scarce and expensive resources in communications, such as the processing capacity of nodes, the bandwidth capacity of transmission lines and buffers at the nodes. All these resources and their management are also valid in communications over the ISDN.

As pointed out above, one of the most important features of ISDN is the provision of multiple channels at the user–network access interface (UNI) and its flexible control by the use of common channel signalling (CCS). It is these two features that are concentrated on in this book, i.e. the use of CCS to control channel access, and dynamic bandwidth variation at the UNI. In this book, the phrase 'channel management' is used to represent a group of related activities and policies relevant to the interface bandwidth resource management. These include bandwidth allocation, bandwidth management and channel assignment and transmission control activities within the UNI. These activities are described by specific policies for the system. The following distinctions are made between them in the ISDN context:

- *Bandwidth allocation*[3] Describes *how* the total bandwidth of a communications facility is distributed (allocated) among different contending users. It is assumed that the bandwidth applies to the whole communications interface.

[3] A bandwidth allocation policy is sometimes referred to as a (bandwidth) access control policy, a channel access policy or a capacity assignment policy.

- *Bandwidth management* Describes the *means* (or procedures) by which the bandwidth of a communications facility is managed. It works in collaboration with, and within the confines of, a bandwidth allocation policy. At the microscopic level, it is assumed to apply to the bandwidth of a single channel of variable capacity.
- *Channel assignment and transmission control* Describes the procedures for channel assignment to incoming requests and de-assignment for cost effectiveness, transmission control for data packet forwarding, and receiving and initiating signalling for link establishment and removal.
- *Channel management* Applies to multiple distinct channels at a communications facility and describes their management. A channel management strategy comprises policies regarding bandwidth allocation, bandwidth management and channel assignment and transmission control operations.

We shall now consider these activities in a little more detail.

A bandwidth allocation policy provides access control to new customers (e.g. users, packets, protocol data units—PDUs) for sharing the interface bandwidth. It also regulates changes in the bandwidth allocation to existing channels. For example, a class of users may not be allowed to access more than a certain share in the link bandwidth.

A bandwidth management policy specifies the method and the algorithm by which the bandwidth of a channel is to be increased or decreased. For example, for a given facility, a hysteresis threshold control may be designed for the service capacity (or bandwidth) control of its channels.

A channel assignment and transmission control policy specifies the conditions, i.e. when to set up a new channel and when to remove an existing one. For example, in the case of a connectionless network layer packet, the policy may specify the setting up of a new channel to a hitherto unconnected destination. Similarly, a channel may be totally removed if it is not used for τ seconds. It will deal with the packet forwarding to channel queues. Initiation of D-channel signalling activity and initiation and ending of packet transmission on any channel is also defined by this policy unit. Lower-layer functions of time slot to channel mapping and free-time slot hunting are also carried out under this policy directive.

Channel management encompasses policies for the above activities. Hence, it includes policies on *how* a channel or a group of channels is to be used, e.g. separately or aggregated together to form a larger capacity channel (i.e. a *superchannel*); *when* to set up new links and remove existing links; *when* and *how* to increase/decrease the existing channel capacity to a destination. it may also include a scheduling (routing) policy if multiple channels with individual queues exist to a given destination. A congestion control policy may also be implemented (e.g., throw packets away if the transmission queue gets longer than a certain level). Finally, it may involve traffic monitoring and statistics gathering operations.

1.5.2 Framework

In this book I propose an architecture for the implementation of channel management and discuss how the functionalities described above are met. I specify the sort of management information needed in order to implement these functionalities in the case of a network layer relay. To this effect, I specify the parameters of interest, the decision metrics and their measurement and the decision mechanisms involved. Queuing and simulation models of this architecture are also developed (see Appendix I). It is shown that after suitable simplifications the resulting nodal queuing system of the network layer relay can be solved analytically using an approximate technique, and that it can be used in the validation of the simulation model (see Appendix II).

The channel management architecture (CMA) is extended to include superchannels. This architecture is then compared with the single-channel CMA. Also, the superchannel formation in the primary-rate ISDN is investigated and several modifications and extensions to the existing CCITT protocols for the D-channel are proposed. In the section on policy performance (Chapter 8), a simulation model is then used to compare the performance of different dynamic bandwidth management policies.

The bandwidth allocation problem is not dealt with any further in this book since this is a very wide field of research on its own right. The bandwidth management problem is covered in Chapters 8, 9 and 10 by concentrating on channels of variable bandwidth at the interface. Hence, by the same token, the need for a bandwidth allocation policy that is used in performance models is reduced to: *assign a new bandwidth unit to a bandwidth management request as long as free capacity is available*. This is acceptable, as no contention is assumed to exist in the simplified model used. Therefore, the interaction of multiple channels at the interface using same or different bandwidth management policies is not studied. Lower-layer functions within the channel assignment and transmission control are also thought to be outside the scope of this book.

1.5.3 Dynamic channel management architecture (DCMA)

In order to achieve dynamic channel management, a generalized approach is needed to identify the necessary modules and units of a software architecture such that it could be applied to different types of ISDN user–network interfaces. Such an architecture would also include events and messages that will be encountered at a UNI. In the absence of such an architecture, this book proposes a simple architecture and identifies the necessary elements. Its main features are described below.

An access node interconnected to ISDN will have a physical interface as well as a software interface. The distribution of functionality within this access node can be various for different nodal architectures. In general, this interface can be represented as shown in Fig. 1.5. The dynamic channel management (DCM) module (composed of one or more processes) is shown to have interaction with the interface access and control process. This module needs to have regular access to the status of the particular ISDN interface it is supposed to manage. Figure 1.6 shows the relationship between the DCM and associated parameters that affect it.

The DCM monitors the ISDN interface status continuously by a report sent to it by each relevant event across the interface as well as within the ISDN interface. These events include:

- *Packet sent* A packet has been sent to the interface to be queued to a channel queue.
- *Packet received* A packet has arrived on any one of the channels.
- *Set-up channel* Set up a new channel to the destination or set-up requested by remote destination.
- *Disconnect channel* Disconnect a channel to the destination or disconnect requested by remote destination.
- *Vary channel bandwidth* Increase/decrease channel bandwidth to a given destination.

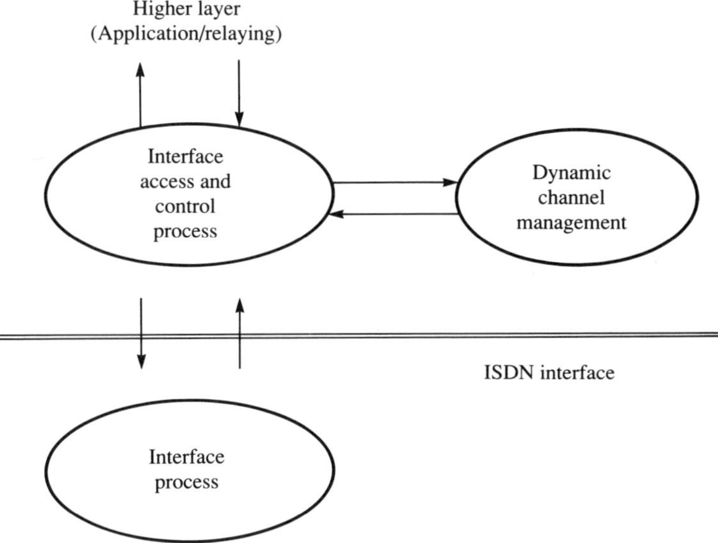

Figure 1.5 A simple schematic of the DCMA

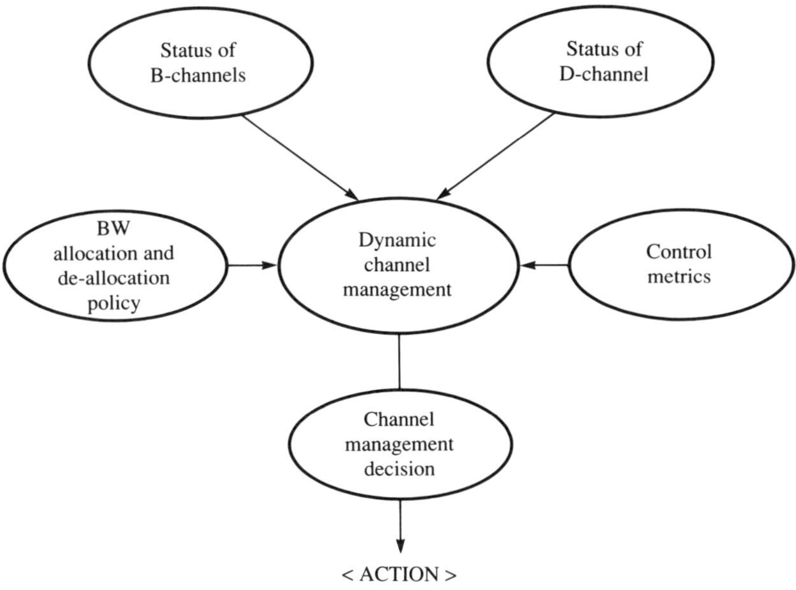

Figure 1.6 Relationship between DCM and associated parameters

Each of these events generates a change in the status of the interface and hence causes an update of status information by the DCM module. Figure 1.7 shows the relationship of the DCM and events sequence within the architecture. As a result of this event, an action command may be generated by the DCM module and sent to the interface process.

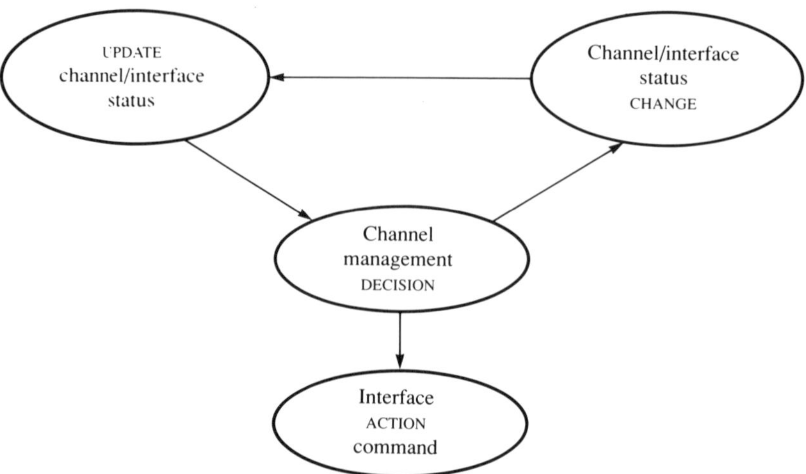

Figure 1.7 DCM events sequence

1.5.4 Superchannel formation

At the user–network interface, ISDN provides multiple fixed-size channels of possibly differing bandwidth capacities (in multiples of 64 kbps). Two problems exist: the provision of $n \times 64$ kbps channels, and the dynamic variation of channel bandwidth. In the absence of the former, an alternative method is the formation of superchannels (channels of arbitrary bandwidth) using the basic channel types (*physical channels*) by the *aggregation* method. Identification and dynamic bandwidth variation of such channels at the UNI needs to be provided (see Chapters 6 and 7). The CCITT Recommendations of the I-series have no provisions for the above. This book addresses the protocol issues relating to this problem (see Chapter 6).

1.5.5 Dynamic bandwidth control

A multi-channel ISDN interface can be organized in a variety of ways, leading to different queuing models. Assuming that no quality-of-service queuing is incorporated, these can be listed as follows:

1. Multiple parallel servers with their individual queues: $n \times$ A/B/1
2. Multiple parallel server groups, each with a single queue: $m \times$ A/B/c
3. Single server groups, each with a single queue but variable service rate: $k \times$ A/B/1
4. A mixture of the above

In queuing models 1 and 2, the variation in the *logical* channel (a channel formed by the association of several physical channels) bandwidth to a given destination is achieved by the addition or deletion of servers and their queues, or just servers, respectively. The type 3 queuing model applied to a single channel assumes the existence of a superchannel structure (see previous section). One immediate question that arises from the formation of a superchannel is *how* to vary the service capacity (rate or bandwidth). Service capacity control, based on a threshold policy applied to the number of customers in the queue or system, has previously been proposed (see Gebhard 1967). Here, the type of the threshold policy becomes important. The effects of policy decisions need to be evaluated. In this book, four queue-based heuristic multi-level hysteresis threshold control policies (see Chapter 8) and three rate-based threshold control policies are presented and evaluated. Queue-based multi-level and bi-level hysteresis threshold control policies (see Chapter 9), as well as queue- and rate-based policies (see Chapter 10), are compared and contrasted.

1.6 REVIEW OF RELATED WORK

A wide area or research is implicated in the topic of this book. The areas of direct concern—the LAN–ISDN interconnection, flexible bandwidth access in ISDNs and the service capacity control of queuing systems—are given wider coverage than the peripheral issues, which are discussed in the later sections.

1.6.1 LAN–ISDN interconnection

The issue of LAN–ISDN interconnection is relatively new, as the emergence of ISDN standards began with CCITT Recommendations of the I-series in 1984. Here, a technical as well as historical review of the LAN–ISDN interconnection issue is presented through the published work of interest researchers.

One of the earliest investigations into the position of LANs in an ISDN world was carried out by Davies (1985), who saw the absence of structured data transmission services in the then emergent ISDN as one of the greatest weaknesses for LAN–ISDN interconnection. The major issues in this interconnection have been identified as data delimiting, message size and flow control. The synergies between the LAN access modes and the passive bus in the basic-rate ISDN are pointed out. In the case of integrated services local networks (ISLNs), offering data, voice and video services operating at very high bit rates, two major issues in LAN–ISDN internetworking are the necessity for higher bandwidths in the wide area network, and the need for the ISDN to be able to transmit (and possibly switch) the packetized voice.

The main reasons for wanting a LAN–ISDN interconnection are listed by DePrycker (1985). The major problems when connecting a LAN to an ISDN have been identified as address inconsistency, frame structure incompatibility, and sub-addressing within a LAN. The address inconsistency is attributed to the use of layer 2 addresses by LANs and layer 3 addresses by the ISDN. Three solutions are offered for address inconsistency. The first involves the level 3 address generation within the LAN by mapping between level 2 and 3 addresses. The second solution involves the use of an adaptor between the LAN and the S-interface to do external mapping (inconsistent with the OSI RM). The third solution proposes the supporting of network layer protocol in each terminal/host connected to the LAN, hence obviating the need for mapping. One way of solving the frame structure incompatibility is to use a gateway to remove all frame structure elements. The solution proposed for the sub-addressing within the LAN is to have the LAN gateway interpret the first in-slot received data and direct the following data to an appropriate address within the LAN. DePrycker also considers the use of X.25 PLP (packet-level protocol) over circuit-switched B-channels and the LAN, where the LAN–ISDN gateway performs the flow control, connection set-up and release. A mapping is to be performed between the X.25 virtual circuit (VC) and the B-channel circuit connection. The disadvantages of this method include the time-consuming VC set-up and the X.25 layer 3 error correction mechanism, considered an over-kill when used over a LAN which has good bit error rate (BER). When connectionless protocols are used over the LAN, the use of X.25 VC 'tunnel' is advocated, where the datagrams are ported using X.25 data packets.

Interconnection of LAN–ISLAN and ISDN have been studied by the author (see Deniz 1985a, 1985b) in general terms. Building wide area networks using the ISDN has been proposed in a paper by Deniz and Matthewson (1986). Here, the problems of interconnecting a packet-switched sub-network to the circuit-switched ISDN are investigated and the CO and CL protocol working over the ISDN B-channels are studied. The circuit set-up and removal operation is defined in both cases. When a CL protocol is used, the circuit is to be established with the first datagram to a given ISDN destination and removed after a connection *time-out* mechanism when no further packets arrive to the gateway for that destination. A further proposal includes the use of multi-link procedures over multiple B-channels giving point-to-point connections of larger bandwidths. The possibility of using the *fast-select* facility of X.25 for transporting datagrams over semi-permanent ISDN lines is also suggested.

The issue of interconnecting LANs, PABXs and ISDNs is also investigated by Zylberberg (1986), who gives a survey of the activities by ECMA in developing standards for the LAN–PBX interworking. This paper also investigates the methods available for LAN–ISDN interconnection. It summarizes the methods as circuit-switched and packet-switched modes. The definition of new ISDN packet modes using full out-of-band signalling and level 2 multiplexing are reported. The use of a terminal adaptor (TA) acting as a gateway between the LAN and ISDN for the first two solutions is advocated. The issue of connecting an ISDN terminal to a LAN is also raised. The suggested solution uses only existing data terminals over LANs and uses a PABX for ISDN-type terminal connection.

One of the earliest projects looking into the LAN–ISDN interconnection issues is the Unison Project (Clark 1986; Clark *et al.* 1986; Tennenhouse *et al.* 1987), sponsored by the UK Alvey directorate. This project had its origins in the Project Universe (Burren 1985; Leslie *et al.* 1984), which was established as a collaborative endeavour to explore the linking of LANs by broadcast satellite and terrestrial link to form a high-speed wide area computer network and to investigate the integration of services. The Unison Project initially used a subset of the Alvey High Speed Network facilities and initially consisted of fixed (plug-patchable) 2 Mbps links and then $n \times 64$ kbps circuits individually switched by an operator at a central console attached to automatic cross connection equipment (ACE) (see Clarke 1986). This switching architecture would not support fast and flexible switching of individual 64 kbps time slots using the D-channel signalling over the time slot 16.

The issue of LAN–ISDN interconnection, and the type of protocols to be used over these connections, the implications for gateway design has been raised in the UCL Annual Report of research activities in October 1986 (see Deniz 1986). In this report, the possibility of using integrated services and multi-media application types, both at the local and wide area network domains, and the impact of this possibility on communication protocols were raised. Building of specific connection-oriented and connectionless mode network layer relays between LANs and ISDN are the topic of papers by Deniz (1987a, 1987b) and Knight *et al.* (1987). In the first of these the 'call management', and in the latter the 'circuit management'[4] phrases are coined for the call set-up and removal operations of the circuit-switched ISDN referred to earlier in Deniz and Matthewson (1986).

Tennenhouse *et al.* (1987) reported that the Unison Project would study the practical aspects of multi-media services and distributed computing across multiple LAN sites interconnected using the standard (G.732) 2.048 Mbps PCM links (known as 'Megastream' in the United Kingdom). The pilot network provided circuit switching through ISDN switches, and, by then, supported the CCITT layer 2 (Q.921) and layer 3 (Q.931) signalling protocols.

Berera and Jardin (1987) describe a method of interconnecting an Ethernet LAN to a PABX using the PRISDN interface given by the ECMA 104 standard (ECMA 1985a) (based on I.431/G.703/G.732 with additional features such as maintenance loops). The first version of their equipment is an Ethernet-based terminal server. It is actually a LAN–PABX gateway that supports the basic functionality required for the support of low-speed asynchronous terminal equipment (TE) connections. It also provides 64 kbps clear connectivity. The D-channel protocol versions of ECMA 105 (1990a)/Q.921 (layer 2) and ECMA 106 (ECMA 1985b) (layer 3) are developed for signalling purposes. Cooper (1987) discusses the influence of ISDNs on local area networks and shows how the LAN must adapt to the network management requirements of the ISDN. It further considers the merging of LANs and PABXs developed to ISDN standards. The LAN–ISDN gateway is said to provide NT2 functions, to correct electrical termination to the ISDN, to transmit and receive information at ISDN rates and with the ISDN protocols, and to provide network management functions. The second requirement implies that (asynchronous) packet-oriented to (synchronous) circuit-oriented conversion and buffering facilities must be provided at the gateway. The third requirement implies that the LAN or the gateway is equipped with a comprehensive network manager capable of monitoring the LAN traffic, determining the acceptance of incoming calls, establishing call priorities and registering details of the properties of all devices connected to the LAN. Henriet (1987) considers the integration of LAN, PABX and the ISDN equipment and services and its benefits to users, but offers no technical proposals.

A study by Kirstein *et al.* (1989) pointed out the possible impact of ISDN on the Joint Academic Network (JANET) in the UK. According to this, the benefits of the ISDN in an academic network environment for the period investigated will be the use of ISDN lines as catastrophe backup, on-demand extra capacity, direct capacity and remote terminal access lines.

[4] Note that this terminology is similar to the 'channel management' terminology adopted in this book (see Chapter 4 for the definition). The latter is considered to be more appropriate, since a channel may actually have multiple 'circuits' associated with it, as indeed may be the case for a superchannel (see Chapter 6), a channel of arbitrary bandwidth in multiples of 64 kbps, which may be aggregated from individual B or H-channels set up separately. Furthermore, since the time slot(s) to channel mapping is left to the user in the ISDN user–network interface, and the bandwidth of a superchannel is assumed to be a variable under control, the term 'channel management' is better suited.

Deniz and Knight (1989a, 1989b) define the channel management in LAN–ISDN gateways (relays) and propose a *status table* approach to channel management. The use of instantaneous or averaged queue length and/or traffic arrival rate as the control metric for channel service capacity is also proposed. The factors affecting the choice of queuing strategies in LAN–ISDN gateways are also studied. A design for a simple Ethernet–ISDN connectionless network layer relay incorporating channel management is presented. In the case of CO protocols (e.g. X.25) over ISDN, multiple virtual circuits over a single B-channel and multi-link procedures over multiple B-channels are proposed. For the case of CL protocols over ISDN, the effect of the transport protocol time-outs are also studied.

A bandwidth management system is proposed in Harita and Leslie (1989), and is used to aggregate time slots in a PRISDN interface to give instantaneous bandwidths of size $n \times 64$ kbps. An algorithm running on transputers and providing an aggregated wide-band circuit (using $n \times$ B-channels of a PRISDN) is presented in a paper by Burren (1989). A fixed-size cell structure is superimposed over this variable bandwidth circuit-switched channel, giving an asynchronous transfer mode (ATM) type performance. Tennenhouse and Leslie (1989) report on the multi-service protocol suite used over this ATM-type network. The protocol suite used is based on the OSI layering principles but does not conform to the OSI in the distribution of layer functions. Two important aspects of this protocol suite are reported: first, the multiplexing is done at the link layer, making application-specific quality-of-service information, available to scheduling components, both in the network and the end systems; second, parallel associations are used to support the out-of-band implementation of many layer functions.

Fan (1989) gives two designs for CO and CL Ethernet–IDA[5] gateways in the absence of an ISDN in the United Kingdom conforming to the CCITT recommendations during the lifetime of that development. The CO gateway uses X.25 over a single B-channel, where multiple virtual circuits can be supported. The LAN–ISDN internetworking is achieved at reference point R (see Chapters 2 and 3). Both of these gateways used simple set-up and removal operations for the single 64 kbps channel available at the BR–IDA interface based on the discussions presented in Deniz and Matthewson (1986), Deniz (1987b) and Knight *et al.* (1987).

INCA (Integrated Network Communications Architecture), a European collaboration under the EC-sponsored ESPRIT-1 (European Strategic Programme of Research in Information Technology) programme (1985–8), has looked into the LAN and WAN interconnection issues. However, it has done little work on the LAN–ISDN interconnection issue (Knight and Kirstein 1987). PROOF (Primary Rate ISDN OSI Office), an ESPRIT-2 project started in 1989, intends to bring together the PRISDN and the OSI-based data communications technology to the development of applications in distributed office systems (Knight 1989). The project will develop a high-performance LAN–PRISDN relay conforming to OSI principles and will provide an environment for the porting of X.400 mail, X.500 directory services and multi-media services over a network of concatenated LANs, private ISDNs and PABXs.

[5] Integrated digital access (IDA) is the pre-ISDN pilot network of British Telecom (BT). It has two variants: the basic-rate IDA and the multi-line IDA, corresponding to the BRISDN and PRISDN of CCITT, respectively. However, the BR-IDA has X.21 at its physical layer and the ML-IDA has a BT proprietary standard called DASS 2 (Digital Access Signalling System, version 2).

The IEEE 802.9 IVD–LAN Interface Working Group considered the LAN–ISDN interconnection via frame relay (IEEE 1988a). The method described is based upon the extended-bridge approach (MAC layer interconnection) and provides transparency while necessitating only minimal modifications to the existing LLC (link layer control protocol used over LANs) and LAP-D (link access protocol over the D-channel) protocols. Walton *et al.* (1991) describe a design for the Ethernet–ISDN connectionless gateway. The ECMA Technical Group 10 (ECMA 1990b) is currently preparing a Technical Report on the LAN–ISDN interworking which will consider the LAN–ISDN as well as LAN–ISDN–LAN interconnection issues using the ISDN bearer services.

1.6.2 Flexible bandwidth management in ISDNs

The issue of flexible bandwidth management in ISDNs is a relatively new topic of interest in the research community. The asynchronous transfer mode (ATM) (Minzer 1989) in broadband ISDNs has this feature as a potential facility. However, in the narrowband ISDN it has been an open issue. Provision of $n \times 64$ kbps service facility in ISDNs will of course solve the problem partially; the actual mechanism of dynamic bandwidth variation and the control doctrine to be used also need to be defined. At present, the availability of this service in the immediate future is still doubtful.

The need for channel management in LAN–ISDN interconnections, when the interworking is achieved at the network layer using connectionless mode protocols as to when to open (set-up) and close (disconnect) channels and how to vary the link bandwidth, has been raised in Deniz and Matthewson (1986), Deniz (1987b) and Knight *et al.* (1987). The need for a variable bandwidth communication link in LAN–LAN interconnection is also pointed out by Tennenhouse *et al.* (1987), when the integration of bulk data transfer, single transaction type traffic based on remote operations, voice and other multi-media services are to be integrated over a wide area connectivity. It is assumed that any two LAN sites should establish a link 'at start of day', and keep it until end of the day, varying the link bandwidth during the day as the traffic load between the two sites necessitates. It is also assumed that all the traffic between a pair of sites will be aggregated together at the source site and transported along a shared channel to the destination site.

The characteristics of services requiring multi-slot connections and their impact on ISDN design are described in a paper by Roberts and Van (1987). Roberts and Liang (1988) study the performance of the X.25 multi-link protocol for $n \times 64$ kbps data transmission in ISDN, while Altarah and Motard (1989) and Boltz *et al.* (1989) study the dynamic usage of bandwidth in narrowband ISDN. The latter two papers propose methods for solving the time slot sequence integrity (TSSI) problem in ISDNs.

Deniz and Knight (1989a, 1989b) provide a simple bandwidth management architecture for dynamic channel management. Harita and Leslie (1989) present analytic and experimental results for dynamic bandwidth management using threshold control. Burren (1989) describes an algorithm for solving the TSSI problem. The protocol issue in channel management have been the subject of the paper by Deniz and Knight (1990). The problem of channel management and dynamic bandwidth management formed on queue-based multi-level threshold control policies and their performance have been the subject of a study by Deniz (1991). A recent study by Harita (1991) also investigates dynamic bandwidth management in an ATM overlay over ISDN. A paper by Altarah and Seret (1991) propose a closed-loop control method for rate-based bandwidth management. The rate-based bandwidth control has also been the subject of study by Du and Knight (1991).

1.6.3 Service capacity control of queuing systems

The classical queuing theory deals mainly with *descriptive* models, where, for specified characteristics of a queuing process, the state probabilities and (expected value) measures of effectiveness describing the system are obtained. In contrast, the control of queuing systems results in a *prescriptive* model. The prescriptive queuing models also fall into a category of queuing theory referred to as the 'design and control of queues' and they prescribe the optimal course of action to follow (Gross and Harris 1985). The design and control models relate to static and dynamic models, respectively. Static models prescribe a combination of control parameter values which optimize certain cost functions. The control models are usually involved in determining the optimal control policy. Queuing processes can be characterized by the following six features (Gross and Harris 1985):

1. Arrival pattern
2. Service pattern
3. Queue discipline
4. System capacity
5. Number of service channels
6. Number of service stages

The control of a queuing process can therefore be exercised by controlling one or more of these characteristics. In this book we are interested only in the service capacity control of a queuing system that results directly from the multi-channel capabilities of an ISDN user–network interface as described in Section 1.4 above and Chapter 2. Service capacity (rate) control in queuing systems using a threshold mechanism superimposed on the instantaneous queue length has been studied in the literature over the past three decades. A general classification of this type of service capacity control is given in Chapter 9. The studies in this field can be divided broadly into two groups: those in which the system service capacity switching (on or off) is assumed to be instantaneous, and those where switching takes place after a (random or fixed) delay.

Instantaneous switching The earliest research in this field is attributed to Romani (1957) and Moder and Phillips (1962) (see also Saaty (1961) and Gross and Harris (1985)). Both of these works study the M/M/c queue where the number of servers is the variable under control. In Romani's control doctrine, if the queue builds up to a certain critical value (*threshold*) N, additional servers are added as new arrivals come. The maximum number of servers available is assumed to be infinite. Hence the queue never exceeds the value N. Servers are removed when they finish service and the queue length is zero.

In the Moder–Phillips (1962) study (see also Chapter 9 and Appendix IV), a minimum number of servers σ ($\sigma \geq 1$), is always available and the maximum number of servers, S, is fixed. The switching-on of additional servers is done in a similar fashion to Romani's model until the limit S is reached. The additional servers are removed when they finish serving and the queue falls below another critical value υ. This work provides closed-form expressions for the mean queue length and mean number in the system, as well as idle time, mean waiting time and the number of interruptions (server starts). This model is studied in more detail in Chapter 9 and we propose a new expression for the mean number in the system (see Appendix IV).

Yadin and Naor (1967) have generalized the work in this field. In their model, customer arrivals are assumed to be Poisson and service exponential. The service station is capable of providing service at different controllable rates, so the queuing model is essentially M/M/1 with state-dependent service. Alternatively, the number of active servers within the station may be varied in some fashion, producing a variable-capacity service station. It is noted that, if the service capacity increases (set-ups) incur costs, it is desirable to reduce the number of service capacity changes per unit time to a minimum. In such a situation, the optimal doctrine may provide for delay of setting up new (or removing active) service capacity, leading to a *hysteresis* phenomenon. The following class of policies is considered. For a sequence of feasible service capacities $\mu = (\mu_0, \mu_1, \ldots, \mu_k, \ldots)$ where $\mu_{k+1} > \mu_k$ and $\mu_0 = 0$, the two integer sequences $\mathbf{R} = (R_1, R_2, \ldots, R_k, \ldots)$, $\mathbf{S} = (S_0, S_1, \ldots, S_k, \ldots)$ such that $R_{k+1} > R_k$, $S_{k+1} > S_k$, $R_{k+1} > S_k$, $S_0 = 0$, represent the management doctrine. The policy is stated as follows: 'Whenever system size reaches a value R_k (from below) and service capacity equals μ_{k-1}, it is increased to μ_k; whenever system size drops to S_k (from above) and service capacity is μ_{k+1}, it is decreased to μ_k.'

An M/M/1 queuing process using a single hysteresis with bi-level service capacity control has been studied by Gebhard (1967). The policies described by Gebhard are special cases of the Yadin–Naor (1967) policies. The control doctrine used prescribes the average service rate depending upon the queue length. A single-threshold model is also analysed for completeness. The terminology used in Gebhard's paper to describe the control doctrines is somewhat different from that adopted in this book (see Chapter 9). In the single- (point) threshold model, the service rate is μ_1 for queue lengths less or equal to N_1 and μ_2 ($\mu_2 > \mu_1$) for queue lengths greater than N_1. In the case of the bi-level hysteretic control, the service rate is increased from μ_1 to μ_2 whenever the queue length increases to a value equal to or greater than an upper control level N_2. The service rate is reduced from μ_2 to μ_1 whenever the queue length drops to a value N_1 ($N_1 < N_2$). For both models, closed-form expressions are obtained for the expected value and variance of the number of customers in the queue. In both cases the steady state is assumed to exist provided the higher of the two service rates, ρ_1, is less than unity. Two *figures of merit* for measure of performance of the bi-level hysteretic control doctrine are derived: the rate of switching between service rates, and the proportion of time that the queuing system operates at the higher service rate.

Several cost formulae are superimposed on the queuing systems considered for a comparison of costs when the control doctrine is varied. Although optimum solutions are not derived, it is concluded that the *optimum* control doctrine depends on the cost formula. The total cost formula is assumed to have two components: cost C_s associated with service (including switching costs) and cost Cq associated with queuing. The single-level control is found not to be best for any cost combination, while the hysteretic control is found to be best under certain cost combinations. The results obtained exemplify that cost allocations do not always make the best control option apparent.

Delayed switching Yadin and Naor (1963) study an M/G/1 queuing system with a removable service station and introduce delays in both the set-up and close-down operations. Their work considers a single server which could be either present or absent (on/off control). The distribution of the switching delays as well as the service times are assumed to be general. They state that, in the absence of any financial (or other) penalties for the establishment and dismantling of a server, the trivially optimal doctrine is one that eliminates the *idle* fraction of the server by closing down the server and reopening it at suitable times. However, ordinarily the dismantling of a server leads to penalties arising from: an increase in average queue length and queuing time, set-up/close-down times and costs. Hence, their study looks at the effect of an $(0, R)$ doctrine ('dismantle the service station, if 0 customers are in the queue; re-establish the service station, if the queue has accumulated to size R') on the above queuing system. Their results include the average queue length and queuing time. A cost structure is also superimposed on the system, and optimization procedures are outlined.

An M/M/2 system with delay before the second server becomes available is introduced by King and Shacham (1986, 1989). The system with delayed second server is also studied by King (1990). In their study, King and Shacham (1986) look at a system with a single queue and two parallel identical servers where one is available permanently and the other is switched in after a random switching delay (exponentially distributed with mean $1/\zeta$) whenever the queue length reaches K. The second server is switched off (instantaneously) only when the queue length falls back to zero. That is, the control doctrine used is a single hysteresis threshold with bi-level capacity control. (See Chapter 9 for a classification of threshold control of queuing systems.) A cost structure composed of two factors—cost of waiting time (queuing time plus service time) and cost of dial-up (channel set-up)—is superimposed on the system. According to this, at light system load the switching cost is dominant, and at loads close to capacity, waiting is the dominant factor in cost. The optimum threshold is found to be a decreasing function of the offered traffic to the queuing system. The cost function used has a low gradient in the region of the optimum threshold. Therefore, it is concluded that choosing a threshold in the vicinity but not exactly at the optimum threshold will not increase the operating costs excessively. It is also found that, for fixed threshold values K, the cost increases with load for large K but has a minimum for small K.

A second model introduced by King and Shacham (1986) deals with two queues with a dial-up channel operating in full duplex mode. Each node that has its buffer occupancy above the threshold dials up a channel. When the first dial-up is complete, the other buffer aborts any ongoing dial-up process and uses the established channel. The channel is released when both buffers are empty. No closed-form results are obtained, but a numerical solution procedure is outlined. Again, it is found that the choice of threshold is not critical as long as the operating threshold is close to the optimal.

Generalization of the above to an M/M/1 system with delayed service capacity increase is carried out by Harita and Leslie (1989). The control doctrines used are single (point) threshold with bi-level capacity control and single hysteresis threshold with bi-level capacity control. The channel switching-on delay is assumed to be exponentially distributed with parameter γ. The switching-off is assumed to be instantaneous. Expressions in z-transform for the mean and variance of the number of customers in the queue are obtained.

1.6.4 Other issues

Standardization bodies active in ISDN The work on ISDN can be reviewed under three groups: the national and international organizations responsible from standards development, the service providers (e.g. PTTs) and manufacturers, and the users. ISDN is such a wide field that it is almost impossible to review all the relevant work in this field. Therefore, here I present the work of national and international organizations, while the other related work, especially research aspects, are referred to in the rest of the book where necessary.

The CCITT (International Telegraph and Telephone Consultative Committee) has been working on the definition of the ISDN concept as well as on interfaces and signalling procedures for the ISDN since before 1980. Its study periods are separated into groups of four years. The first ISDN study period, 1980–4, has culminated in the I-series Recommendations known as the Red Book, while the study period 1984–8 has produced the Blue Book version of the recommendations. The current status of these standards, their technical overview and future trends are further covered in Chapter 2. The ISO (International Standards Organization) prepares standards for data communication. CCITT and ISO co-operate in some of these areas. ISO standards are relevant for data communications using the ISDN. Indeed, the layering principle of the Open Systems Interconnection (OSI) is adopted by the CCITT in developing the I-series Recommendations.

In Europe, the European Computer Manufacturers' Association (ECMA) deals with various aspects of data communications between data processing systems, including their interconnection to ISDN PABXs. ECMA has a series of Technical Standards relating to ISDN systems. Also, a new ISDN users' forum, the European ISDN Users' Forum, has been established to co-ordinate the activities of the service providers, manufacturers and users on ISDN services and equipment.

The European Telecommunications Standards Institute (ETSI) was formed by the CEPT (European PTTs) in November 1988 to carry out the standardization activities for the European countries (see Gagliardi 1989). This standardization activity also applies to the ISDN in selecting common options and services. An additional objective is the co-ordination of the installation planning activities. A group has also been formed within CEPT to study the use of ISDN to provide the OSI network service.

QUAD is joint venture between the German, French and Italian PTTs and British Telecom, which are co-operating within this venture to produce documents setting out how the ISDNs will be implemented. Its aim was to achieve terminal portability through technical harmonization by 1992 (Trudgett 1988).

The Standards Promotion and Application Group (SPAG) is a European industry grouping formed to promote functional standards and a common conformance testing facility (see Trudgett 1988). SPAG produces a Guide to the Use of Standards (GUS) which contains profiles as well as a description of their use and purpose.

In the United Kingdom, the UK ISDN Standards Forum (UISF) comprises UK service providers (e.g. British Telecom) and users; it co-ordinates the submission of information and views of the UK organizations to the CCITT on relevant standards. The Department of Trade and Industry has also undertaken work to co-ordinate ISDN standards, implementations and usage in the UK. Another forum that is co-ordinated by the British Standards Institute (BSI) is the TCT/8/-/4 Committee, looking at the LAN–ISDN and LAN–ISPBX interconnection issues. The work of this forum is also passed on to the relevant committees of ECMA.

In the United States, the Exchange Carriers Standards Committee (ECSA) sponsors the Standards Committee T1—Telecommunications. The membership of the committee is open to all interested parties. T1 develops national standards which are promulgated by the American National Standards Institute (ANSI). It also co-ordinates the US contributions to

CCITT. The T1S1 subcommittee of T1 deals with the ISDN standards in the United States. The North American ISDN Users' Forum (NIU-Forum), which is set up by the National Institute of Standards and Technology (NIST) to look into ways of using ISDN and developing new applications and services, supports two sets of committees: a user set and an implementors' set.

Another organisation working on standards in the United States is the Institute of Electrical and Electronic Engineers (IEEE). The relevant technical committees of this institute have been working on the use of frame relaying techniques in LAN–ISDN–LAN interconnection (IEEE 1988a) as well as B-ISDN and interworking with metropolitan area networks (MANs).

Integrated switching The requirements of different types of traffic sources have traditionally been addressed by the circuit-switched (CS) and packet-switched (PS) communication systems. For example, the voice and data services are better served by circuit and packet switching, respectively. In order to improve bandwidth utilization and reduce the multiplicity of access and network arrangements necessitated, transmission of circuit-switched and packet-switched traffic sources on the same medium can be achieved by using time division multiplexing (TDM). A time division multiplex frame can be separated into two parts: one for the circuit-switched applications (blockable traffic type, class 1) and one for the packet-switched applications (queuable traffic type, class 2). The boundary between the two can be either fixed or movable. The movable-boundary strategy was first proposed (see Schwartz 1987), by Kummerle (1974) and Zafiropulo (1974). The method is shown to offer a better performance advantage for class-2 users over a fixed-boundary scheme, in which the same reserved slot allocations are made. The blocking performance of class-1 calls remains the same in either case. The data packets achieve a reduction in their queuing delays under the movable-boundary strategy (Schwartz 1987). Weinstein *et al.* (1980) have shown that a dynamically movable boundary is not very effective in allowing class 2 traffic (e.g. data packets) to utilize free capacity arising from momentary fluctuations in class 1 traffic (e.g. voice) unless an appropriate flow control mechanism is used. The analysis and design of hybrid switching networks is carried out by Gitman *et al.* (1981), among others. Maglaris and Schwartz (1982) proved that the movable boundary provides optimal or near optimal system performance for the single-user case. The multiple-user case has been analysed by Kraimeche (1984), who showed that for the balanced traffic case, and for the range of parameters used, the movable boundary mechanism provided an improvement in packet traffic delay.

A refined hybrid switching method, the slotted envelope network (SENET) concept is proposed by Coviello and Vena (1975). Here, pseudo-synchronous 'envelopes' are used to convey multiplexed traffic through the network using a wideband slotted transmission format, such that the traditional characteristics of both circuit and packet switchings are simultaneously provided. This concept can provide flexibility for a varying mix of traffic and the capability to provide variable channel bandwidths to match different applications.

Access control problem in ISDNs Narrowband ISDN can be viewed as a time division multiplex (TDM) facility where each time slot in the TDM frame can be accessed separately. When larger bandwidth links are required, multiple slots are accessed. A mixture of traffic comprising K types of customers is assumed to share this facility. Such a situation occurs when more than one class of traffic (e.g. voice and data) are allowed to access the facility.

For a hybrid system with packet- and circuit-switched traffic types, the access control[6] problem can then be defined as: 'dynamic assignment of bandwidth units (slots, bits, etc.) to incoming requests, as a function of both circuit (CS) and packet (PS) switched load, in order to optimise the performance of PS traffic to achieve minimum delay given an upper bound of acceptable blocking probability for the CS traffic' (Kraimeche 1984). A considerable amount of interest has been shown in the performance of the above model in the last eight years. Hybrid switching in TDM systems has been studied by Maglaris and Schwartz (1979, 1982), Li and Mark (1984), and Yum and Schwartz (1987). A taxonomy for hybrid switching is given by Goeldner (1985). Modelling and simulation techniques are used to obtain message, packet or bit-level performances of the models described. However, only a few have concentrated on the message-level analysis (Daigle and Whitehead 1984). A good survey of previous work in this field is given in Fletcher *et al.* (1986).

Channel access control strategies are needed for an integrated communication system with heterogeneous users (data, voice, video) who have differing service requirements, so that it can handle its traffic demands. Close control of access and switching at the input node is necessary for high efficiency and flexibility (Kraimeche and Schwartz 1984). The same applies if ISDN is used in a data-only environment, for example LAN–LAN interconnection using the ISDN. The difference is that, although the real-time requirements of voice and video do not exist, the requirements of various data services still need to be addressed (Deniz 1987b). For example, a suitable quality of service must be provided to users, requesting terminal access in order to provide a workable environment. Services with strict delay requirements may be given higher priorities over those with less stringent requirements (e.g. file transfer and electronic mail).

In the remainder of this section, a brief consideration is given to the queuing models applicable to ISDN access and for the corresponding data-only application model for CL packet access over CS ISDN.

Analytic models for ISDN access Queuing systems for ISDN access have generally been modelled as multi-server queuing systems with F identical, parallel servers, and where customers upon arrival demand a variable (random) number of servers (basic bandwidth units — BBUs) for the duration of their service times. The random number of servers required represents customers' bit rate requirements. The most important property of this class of queues is that a given customer can enter service only when all of his or her requested servers are available.

Different customers are identified by their bit rate (bandwidth) requirements. Other attributes characterizing customer traffic features are the average service (or holding) time, and their generation rate and pattern. A customer arriving at the ISDN access node makes an *access request.* Access requests for bandwidth service are assumed, in general, to be of K types. A type i access is identified by its arrival rate λ_i in requests/time, its mean service time $1/\mu_i$ in units of time, and its information bit rate b_i in BBUs. Two variants of this model exist. In the first, the servers allocated to the same customer are all released simultaneously. This is the model used by Brill and Green (1984) and by Kraimeche (1984). In the second model, the servers allocated to the same customer free up independently. This model has been studied by Green (1980, 1981).

[6] Access control is not related to authorization and security aspects as defined here. It is used to mean permission control to bandwidth resource sharing. The problem is also known as the 'bandwidth allocation problem'.

In hybrid switching models, with both the CS and PS traffic access, the CS traffic is usually assumed to operate on the 'blocked calls cleared' principle and the PS traffic is assumed to be queuable. The same principle is extended to analytic models where *K* traffic streams compete for the bandwidth resource so that two variants of the model are studied: the case where all *K* traffic streams are assumed to be blockable or queuable, respectively. A special case of this model is when the traffic types are categorized into two classes: wide-band (WB) and narrow-band (NB) traffic types. In this case either one or both are assumed blockable or queuable.

Channel assignment techniques in hybrid switching are analysed by Schwartz and Kraimeche (1982). The performance of blockable WB and queuable NB traffic types is studied by Kraimeche and Schwartz (1985b) as a mixed blocked/queued traffic. The case of blocked calls cleared for NB and WB traffic with trunk reservation is studied by Liao *et al.* (1989). A multi-server queuing system with blockable NB and queuable WB customers under a wide-band restricted-access strategy is studied by Serres and Mason (1988). The case of *K* heterogeneous CS traffic types in the 'blocked calls cleared' model is studied by Kraimeche and Schwartz (1984). The performance of complete sharing and complete partitioning is investigated. A class of restricted access policies is also proposed with improved performance. The case of all queuable *K* streams of traffic is considered by Kraimeche and Schwartz (1985a, 1986). A channel access structure for wide-band ISDN is studied by Kraimeche and Schwartz (1987a, 1987b), where the interface is ordered hierarchically in channel capacity and *K* queuable traffic types are allowed to contend for the bandwidth with channel overflow facility. The case of *K* traffic types grouped into clusters with blocked calls cleared and a restricted access policy is studied by Kashper (1987).

Queuing model for CL mode packet access to CS ISDN For the CL packet access to the circuit-switched ISDN, the general queuing models mentioned above are no longer valid (see Deniz and Knight 1989a). In the first place, the concept of customers *demanding* a (random) number of servers is not true for this case. This is because the customers arrive as disjoint packets (datagrams) and have no fields in their headers to indicate the bandwidth that will be required for a complete session (ISO 1987b; Postel 1981a). In fact, they do not have a means of specifying the bandwidth required for that particular packet, either. Second, each customer can enter service when at least one BBU (channel) is available. However, it will not be efficient to assign a new channel (e.g. a B-channel of 64 kbps capacity) to every packet source with a different internet address, since some of the internet addresses will inevitably be on the same subnetwork (e.g. hosts on a LAN) and hence will be served by the same ISDN gateway. In such cases, the traffic to the same subnet can be multiplexed over one or more channels or a superchannel of variable bandwidth.

Multiplexing all the different data traffic sources (e.g. file transfer, e-mail and terminal access) to the same ISDN destination (e.g. a gateway) may produce problems in the provision of quality of service (QoS). It is necessary to study the queuing systems for QoS in CL interworking in order to see the effect of such multiplexing and bandwidth control algorithms on the performance of different traffic types. Prue and Postel (1988) suggest separate queues for different classes of data packets and apportion the server capacity accordingly between the separate queues (single server, multiple queues). Further increases in bandwidth of the channel is assumed to occur under the control of a channel or bandwidth management unit. This unit increases bandwidth of a channel by adding further BBUs to the channel capacity on a need basis. Mechanisms for achieving this are the subject of this book. Furthermore, the release of allocated BBUs for a CL customer is done independently rather than simultaneously.

Hence it is necessary to clarify the notion of a 'customer' and the method of bandwidth allocation and release mechanisms within such an environment. In real computer networks, users communicate via processes. Processes communicate with each other to provide services. A single host or application can have multiple processes which are active at one time. Each user process uses a transport level process to provide end-to-end communications on an inter-network or intra-network domain. On a local network a multiplicity of hosts, each supporting a number of such processes, will want to communicate with peer entities at remote sites. At the ISDN access node (e.g. a network layer relay), all communications directed to a specific destination could be viewed as 'one' customer, despite the fact that their source host or processes could be different. The difference in this case is that we now have an 'aggregate customer' whose features of transmission in terms of packet arrival rate, mean service time and information bit rate are changing over time (Deniz and Knight 1989b). In this case, assuming that all arrivals are independently Poisson,[7] the above characteristics could be viewed as discrete random variables with the following distributions:

$$\Pr(b=b_i) = \frac{\lambda_i}{\sum_{j=1}^{K}\lambda_j} \quad \text{and} \quad \Pr(\mu=\mu_i) = \frac{\mu_i}{\sum_{j=1}^{K}\mu_j} \qquad (1.1)$$

1.7 OUTLINE OF THE BOOK

This book is organized into ten chapters. I shall now give a brief overview of the rest of the book.

In Chapter 2 a description of the ISDN and its features relevant to data communications are presented. The relevant features of ISDN in LAN–ISDN interconnection, superchannel formation and tariffing issues are also introduced here. A guide to the present status and future development of ISDN is given at the end of the chapter. This chapter is part of the broad background that needs to be covered by a newcomer to the world of ISDN.

Chapter 3 first presents an overview of the general data network interconnection issues. The issues pertinent to general ISDN and LAN–ISDN interconnection are then investigated. Particular attention is given to the LAN–ISDN relays and the issues arising from their use.

Chapter 4 introduces the concept of channel management in multi-channel network layer relays. The main issues of channel management (CM) are investigated with reference to the modes of communication over the ISDN. The types of management information needed for dynamic channel management are detailed and the control metrics determined. Implications of CM in LAN–ISDN network layer relays are also studied.

In Chapter 5, a design for a flexible dynamic channel management architecture (DCMA) for connectionless network layer relays is provided and reference to an Ethernet–PRISDN relay is made. Variations to the architecture are also presented.

[7] A further assumption for the traffic arriving from different sources over an Ethernet to an Ethernet–ISDN relay is that no collisions are occurring, since this would modify the traffic arrival distribution patterns considerably.

Chapter 6 concerns a study of the superchannels and their protocols. The problems in superchannel formation, identification and dynamic bandwidth variation in ISDN user-network interfaces are very current issues in the ISDN standardization. A protocol solution to this problem is presented with particular reference to the PRISDN based on the user–user signalling facility.

Chapter 7 provides an insight to the methods developed for the solution of the time slot sequence integrity (TSSI) in TDM-based networks. When individual time slots of a TDM multiplex are required to be aggregated to form a larger bandwidth channel, a problem arises in that each time slot may be switched through different switches and paths in the network. This causes a bit ordering mis-sequence which is called the time sequence integrity problem. Various solutions exist for this type of loss of synchronization and these are presented here.

In Chapter 8, the queue-based bandwidth management technique is investigated through the use of simulation modelling of a connectionless network layer relay. Four heuristic bandwidth management policies are studied and their performances analysed.

In Chapter 9, a comparison of different service rate control doctrines pertinent to the bandwidth management issue is carried out. This chapter also deals with the practicalities of the queue-based bandwidth control doctrines.

Chapter 10 examines the rate-based bandwidth management. The rate-based counterpart of the queue-based bandwidth control is an important technique that is a practical proposition. Three different techniques of rate-based control are presented in this final chapter.

Analytic and simulation modelling of a communications interface is an important activity in the performance engineering of data networks. A modelling process for the gateway-ISDN modelling is detailed for readers unfamiliar with these techniques in Appendix I.

Appendix II presents a detailed simulator validation to give an insight to readers of how simulation models of communications devices and networks are validated and utilized.

This book concentrates on the circuit-switched (CS) bearer services of the ISDN, and as such a sample call control signalling is given for a complete CS call in Appendix III.

Multi-level service rate control forms a major group of policies in the queue-based bandwidth control. Some results based on the queuing theory are presented in Appendix IV.

Appendix V presents a heuristic algorithm for the delayed channel close (remove) policy.

Appendix VI contains a list of acronyms and abbreviations used in this book.

Finally, Appendix VII lists the CCITT Recommendations relevant to ISDNs.

CHAPTER

TWO

THE INTEGRATED SERVICES DIGITAL NETWORK

This chapter highlights the most relevant features of the integrated services digital network (ISDN) in its application to data communications and the interconnection of local area networks using the ISDN. It also provides a general background to the ISDN and its current development status.

The ISDN has evolved from a digital network concept, and this evolution is expected to continue for years to come (one or more decades). In this respect, the features of ISDN can be discussed meaningfully only with reference to the global principles, concepts and features of the network and its services, or to its *current* consolidated and standardized features, or to proposals for the forthcoming CCITT study period of four years. Beyond that time scale, concrete evaluation can not be made easily owing to the evolutionary process; only migratory paths could be indicated. Here, we shall concentrate mainly on the first two cases, with some reference to the immediate proposals expected for the next study period. Some possible future trends will be indicated.

Since its conception, the ISDN has diversified into two main streams: ISDN (also termed narrowband or N-ISDN in this book) and broadband ISDN (B-ISDN). N-ISDN refers to network access at bit rates up to and including 2 megabits per second (Mbps), while B-ISDN refers to network access at bit rates (up to and) greater than 2 Mbps, usually in tens or hundreds of megabits per second. This book is concerned mainly with the narrowband ISDN. It also concentrates on the circuit mode bearer service, because of its wider and earlier availability in Europe. However, a summary look into other aspects of N-ISDN as well as at the current and expected developments in B-ISDN are also presented here for completeness. Furthermore, this chapter refers mainly to the most recent CCITT Recommendations of the I-series, the Blue Book (CCITT 1988a), and its predecessor known as the Red Book (CCITT 1984). Designated work areas for the third study period of 1988–92 are also indicated where necessary. The existing CCITT Recommendations of the I-series and other series relevant to ISDN are listed in Appendix VII.

In the rest of this chapter, the principles and technical features of ISDN are discussed in Sections 2.1 and 2.2. Then, data communication, packet modes and packet network interworking in ISDN are presented in Sections 2.3 and 2.4. In Section 2.5 protocols and their standardization are treated. A discussion on B-ISDN follows in Section 2.6. Section 2.7, on the current status and future trends, is intended to show the developments achieved so far and expected future directions in ISDNs. Relevant features of ISDN for LAN–ISDN–LAN interconnections are then examined in Section 2.8. An introduction to aggregated channel structures is presented in Section 2.9 as a prelude to later discussion on the superchannels (channels of arbitrary bandwidth) in Chapter 6. In Section 2.10 the tariffing issues in ISDN are considered briefly. Finally, a summary of the salient points is presented.

2.1 THE ISDN ERA

The CCITT is the international forum that is leading the way in establishing a world-wide definition and standardization of ISDN. Its studies on the ISDN are published as Recommendations (Rec. for short) of I-series. The I-series Recommendations cover all aspects of ISDN under the following sub-sections: I.100-series, General Concept and Structure; I.200-series Service aspects; I.300-series, Network aspects; I.400-series, User–network interface aspects; I.500-series, Internetwork interfaces; I.600-series, Maintenance principles. These and other relevant recommendations are referred to in the following sections.

2.1.1 Principles of ISDN

An ISDN can be characterized by its three essential features:

- End-to-end digital connectivity
- Multi-service capability (voice, data, video, and multi-media)
- Standard terminal interfaces

The ISDN concept includes the support of a wide range of voice and non-voice applications in the same network. The service integration for an ISDN will be provided by a range of services using a limited set of connection types and multi-purpose user–network interface arrangements. ISDNs will support a variety of applications including both switched (circuit- or packet-switched) and non-switched connections. They will contain intelligence for the provision of service features, maintenance and network management functions. The specification of access to an ISDN should be by a layered protocol structure. ISDNs may be implemented in a variety of configurations to be determined by the national service providers. Hence the details of the underlying network will not be important for the users, since they will see a common service across ISDNs (save for optional services and features).

2.1.2 Evolution of ISDN

ISDNs will be based on the concepts developed for the IDN[1] and will progressively incorporate additional functions and features at present provided by separate networks, as well as other new ISDN-specific features and services. During the transition period, arrangements will be developed for the interworking of services on ISDNs and services on other networks.

2.2 TECHNICAL FEATURES

Three key ISDN communication capabilities could be used to define the user's perception of an 'ISDN service': the provision of end-to-end digital connectivity between end users, the provision of multiple-access channels; and intelligent out-band (common channel) signalling and user control. The end-to-end digital connectivity will provide larger bandwidths, better signal qualities and better error rates than most of the present networks. Multiple access channels will provide more flexibility in the use of the total bandwidth available at the user–network interfaces; to this end the CCITT has defined different channel types and interface structures. The new common channel signalling scheme has been chosen so as to allow the user to control both the network connections and the delivery of user services and applications independently of the user information paths in an effective manner.

The basic architectural model of an ISDN as described in Rec. I.324, the ISDN Network Architecture (Blue Book) (CCITT 1988a), includes the following main switching and signalling functional capabilities: ISDN local functional capabilities (e.g. user–user and user–network signalling, charging); 64 kbps circuit switched and non-switched capabilities; rates greater than 64 kbps switched and non-switched capabilities; packet-switched and non-switched functional capabilities; and common channel signalling (user-exchange and inter-exchange) functional capabilities. For the circuit-switched connections, user-bit rates less than 64 kbps are rate-adapted as described in Rec. I.460 before any switching can take place in the ISDN. Multiple information streams from a given user may be multiplexed together in the same B-channel, but for circuit switching an entire B-channel will be switched to a single user–network interface. Packet-switching capabilities are formed by two functional groupings; packet handling functional grouping and interworking functional grouping. The first grouping contains functions relating to the handling of packet calls within the ISDN, while the latter ensures interworking between ISDN and packet-switched data networks (PSDNs). Another feature of the ISDNs will be its fast switching ability, made possible by digital switching techniques. Although the performance issues relating to this have not been fully defined, average national circuit switching speeds are expected to be below 1 s while for international calls the aim is for a speed below 5 s. Some objective delay values are proposed in the Blue Book and these are described below.

[1] Integrated digital network (IDN) is an all-digital network which itself has evolved from the analogue telephone network and provides digital transmission and switching (i.e. *integrated*). A major component of this evolution is the use of CCITT Signalling System No. 7 (Appenzeller 1986; Mitra and Usiskin 1991; Schlanger 1986) to provide inter-exchange common channel signalling. The SSN.7 is being enhanced to provide functionalities for the ISDN.

2.2.1 Service aspects

Services provided by ISDNs have been divided into two broad categories: bearer services and teleservices. The bearer service is a type of telecommunications service that provides the capability for the transmission of signals between user–network interfaces and as such conveys information between users in real-time and without alteration to the information 'message'. The teleservice is a type of telecommunication service that provides the complete capability, including terminal equipment functions, for communication between users according to protocols established by agreement between administrations.

Bearer services Two types of bearer services have been defined: circuit and packet modes. The circuit mode bearer service has been defined for 64 kbps, 384 kbps, 1536 kbps and 1920 kbps rates (Rec. I.231, Blue Book: CCITT 1988a). In circuit mode, the 'unrestricted information transfer' with '8 kHz structural integrity' capability is of interest for data applications since it provides switched 'bit pipes' over which transfer of user information can be achieved without alteration (within the quality-of-service (QoS) limitations).

The packet mode bearer service involves packet handling functions. The following service categories identified in Rec. I.230 (Blue Book: CCITT 1988a) are described in Rec. I.232: virtual call and permanent virtual circuit (connection-oriented (CO) mode), connectionless (CL) and user signalling packet mode bearer services. Further information on the last two may be found in the Blue Book. In the CO mode of operation, 'unrestricted information transfer capability' and 'service data unit structural integrity' are supported. User information is conveyed over a virtual circuit (VC) within a B- or D-channel. Signalling may be provided via D-channel and/or VC within B-channel. Access protocols are as specified in Rec. I.440, I.450/1, I.462 (X.31) and X.25 (layers 2 and 3) (CCITT 1988a) (See also Section 2.4).

Teleservices Teleservices use bearer services to move information around, and in addition employ 'higher-layer functions' corresponding to layers 4–7 of OSI-RM. The Blue Book specifies the following teleservices, which are described in Rec. I.240 and discussed in I.241: telephony, teletex, telefax 4, mixed-mode, videotex and telex (see CCITT 1988a for further details). The mixed-mode teleservice provides combined text and facsimile communication.

2.2.2 Network aspects

These cover the network functional principles, the ISDN Protocol Reference Model (PRM), numbering, addressing and routing principles, connection types and performance objectives. The ISDN-PRM, which is based on the principles of OSI-RM, allocates functions in a modular fashion that facilitates the definition of telecommunications protocols and standards. The ISDN-PRM extends the OSI model beyond the traditional point-to-point, user–network–user, in-band signalling, data-only communications. This extension includes the separation of signalling and management operations from the flow of application information within a piece of equipment; definition of communication contexts which may operate independently from each other; and the application of the above to internal network components. This separation leads to the definition of independent communications contexts. Each context can be modelled individually and use independent protocols.

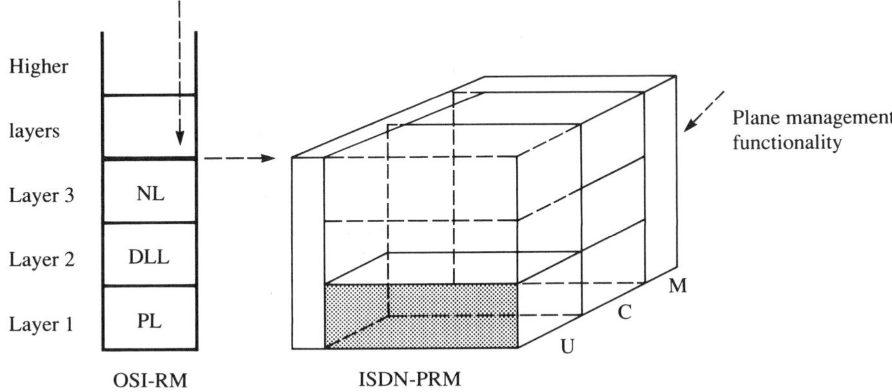

NL: network layer; DLL: data link layer; PL: physical layer
U: user plane; C: control plane; M: management plane

Figure 2.1 OSI reference model (OSI–RM) and ISDN protocol reference model (ISDN–PRM)

The ISDN-PRM refers to two distinct logical planes: user (U) and control (C) planes. The layering principles apply to each of these planes; each one can potentially accommodate a seven-layer protocol stack (CCITT 1988a; Duc and Chew 1985; Potter 1985). A plane management function is required to allow co-ordination between the activities in different planes. Further study is needed to specify the types of layer services required to describe a telecommunications service, data flow modelling and ISDN management. The relationship between the OSI-RM and the ISDN-PRM is shown in Fig. 2.1

Numbering principles within the ISDN (Rec. I.330) will be based on the expansion of the international PSTN numbering plan (Rec. I.331/E.164). An ISDN address is composed of an international ISDN number and an ISDN subaddress of up to 15 decimal digits and 40 decimal digits (20 octets) respectively (CCITT 1988a). The ISDN address may be of variable length, and all ISDNs will be capable of conveying the ISDN subaddress transparently. This subaddress may be used by private communications facilities such as LANs and PABXs or by other private networking arrangements. In order to identify different OSI network layer service access points (NSAPs), methods have been defined in the Rec. I.330 (CCITT 1988a) to relate the ISDN number to the OSI network layer address.

The performance objectives regarding propagation delays, availability and errors were not dealt with in full during the 1980–4 period. These have been extended further in the Blue Book. However, the specification is still by no means complete. Currently, provisional performance values are quoted, while the actual target values are left for further study. The general network performance (QoS) measures are defined in Rec. G.106 (Red Book), while ISDN performance objectives are defined in Rec. I.350 (Blue Book). The network performance objectives for connection processing delays in an ISDN are given in Rec. I.352. These are further discussed below. The network performance objectives regarding error and slip rates that apply at reference point T of an ISDN are given in Recs. G.821/2.

The CCITT G-series Recommendations define hypothetical reference connections (HRX) in order to develop performance objectives for different HRXs. An HRX represents a typical worst-case end-to-end digital connection. The longest (international) terrestrial connection envisaged by CCITT is 27 500 km (Rec. G.801). This represents a *propagation delay* of less than 100 ms. If a satellite link is present, this would easily add up a round-trip delay of approximately 520–560 ms (Peel 1989; Sastry 1984). Recommendation G.821 defines overall performance for HRX at 64 kbps and an end-to-end BER of 10^{-6} on a circuit-switched 27 500 km connection.

Table 2.1 shows the delay performances for set-up, alerting, disconnect and release delays (Rec. I.352) in circuit-switched connections (voice or data) across the worst-case HRX connection (27 500 km) specified in Rec. G.801. The values observed for the first three parameters will be dominated by the number of exchanges in a connection. For shorter-length connections the observed values will be lower. Delays are specified for a nominal busy hour, and the delays that are dependent upon a user equipment (or network) and user response time are not included. Additional delays may be incurred at signalling message queues. Furthermore, the specified performances are for connections exclusively over ISDNs (i.e. no interworking).

Table 2.1 Objectives for connection processing delays in an ISDN (worst-case HRX connection)

Statistic	Delay (ms)			
	Set-up	Alerting	Disconnect	Release
Mean	4500	4500	2700	300
95%	8350	8350	4700	850

2.2.3 User–network interfaces

An important aspect of service integration for an ISDN is the provision of a limited set of standard multi-purpose user–network interfaces (UNIs). An ISDN is recognized by the service characteristics available through user–network interfaces, rather than by its internal architecture, configuration or technology. Hence the CCITT has put most of its initial effort into the specification of this interface to cover both the existing terminal/user base and the future ISDN terminals/users.

The I.400 series of Recommendations (CCITT 1988a) describe the interface between user equipment and ISDNs. They include a reference configuration, giving the functional model for the subscriber, a definition of interface structures and access capabilities, and detailed specifications of layers 1–3 for the basic and primary-rate interface structures. Several other Recommendations in the series provide standardized methods for rate-adapting classical X.1 and V.5 data rates into B-channels, for submultiplexing B-channels and corresponding circuit-mode bearer services, and for interworking with digital network facilities providing only restricted 64 kbps information transfer rates. Furthermore, a terminal adaptor for converting X.21 circuit-switched interfaces into the ISDN basic user–network interface is defined in

Rec. I.461/X.30. Recommendation I.462/X.31 defines how existing packet-switched public data network (PSPDN) services can be provided in or through an ISDN to a packet mode user terminal. Packet mode user terminals, which use the X.25 packet-layer protocol but attach to an ISDN user–network interface, are also covered (see also Section 2.3).

Recommendation Q.940 (CCITT 1988a) defines the ISDN UNI Protocol for Management (General Aspects). The management functions so far identified include: fault, configuration, accounting, performance and security aspects (CCITT 1988a; Ishii 1988).

Reference configurations The user–network interface lies within the customer premises. The reference configurations include reference points which are the conceptual points dividing functional groups. These functional groups include the termination (network termination—NT1), distribution and switching (NT2), terminal equipment with ISDN user–network interfaces (TE1), and terminal equipment with other interfaces (TE2) and associated terminal adaptors (TA). Figure 2.2 shows the reference configurations for the ISDN UNI (Rec. I.411). The use of S and T reference points will be discussed in Sections 3.2 and 3.3.

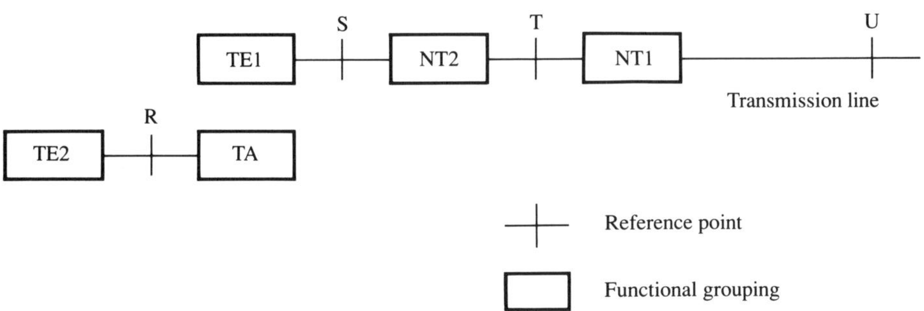

Figure 2.2 Reference configurations for ISDN user–network Interfaces

Channel types A limited set of channel types is defined, including B, D and H channels (Rec. I.412). The B-channel (64 kbps) carries user information. The D-channel (16 or 64 kbps for basic and primary rate, respectively) carries packetized signalling (s-data) information to control the set-up, modification and clearing of calls and services. Surplus bandwidth capacity on this channel may also be used for packet mode user data (p-data) and telemetry (t-data). The H-channels—384 (H_0), 1536 (H_{11}) or 1920 (H_{12}) kbps—carry circuit or packet mode user information. New channel types for B-ISDN are also defined in the Blue Book (CCITT 1988a). These are discussed in Section 2.6.2.

Interface structures Only two general (narrowband) ISDN user–network interfaces are described. These are the basic rate (Rec. I.420) and the primary rate (Rec. I.421) interface structures.

2.2.4 Basic-rate access (BRA)

This interface has an information-carrying capacity of 144 kbps. It is organized as (2B + D) channel types. With the framing, synchronization and D-channel echo bits, the aggregate transmission rate is 192 kbps. This interface is mainly intended for the connection of ISDN terminals or other individual terminals/workstations which will not need very high bandwidths. At this interface both the point-to-point and bus (S-interface bus, or S-bus) distribution types are available. In the S-bus configuration, up to eight terminals can be connected to the NT2, all sharing the same D-channel. Recommendation I.420 provides a contention resolution procedure for B-channel allocation since only two terminals at a time can access these.

2.2.5 Primary-rate access (PRA)

This interface has two subdivisions with differing information-carrying capacities. These are 1536 and 1984 kbps, and they represent requirements of US and European networks. Their aggregate transmission rates are 1544 and 2048 kbps, respectively. This interface can be organized in several ways:

1. Primary-rate B-channel interface structures (23B + D) and (30B + D)
2. Primary-rate H-channel interface structures:
   (a) H_0 channel structures, e.g.: $4H_0$, $(3H_0 + D)$, $(5H_0 + D)$
   (b) H_1 channel structures, e.g.: H_{11}, $(H_{12} + D)$
3. Primary-rate interface structures for mixtures of B and H_0-channels:
   (a) Can consist of a single D-channel and any mixture of B and H_0-channels
   (b) Example structures: $(5H_0 + D)$ and $(2H_0 + 11B + D)$

It is interesting to note that, if a D-channel in one primary-rate interface structure is not activated, the D-channel of a neighbouring interface structure may be used for carrying signalling information for that interface structure.

The physical layer of a primary-rate interface conforms to CCITT Recs. G.703/4/6. Figure 2.3 shows a time division multiplex (TDM) frame structure (Rec. I.431/G.704) with associated time slots (TSs).

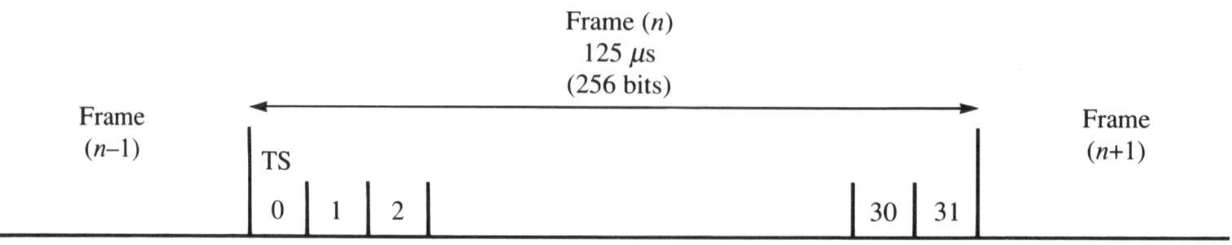

Figure 2.3 TDM frame structure used at PRA physical layer (Europe)

2.3 DATA COMMUNICATION IN ISDN

Data communication in ISDN can be achieved by using either the circuit- or packet-switched and/or non-switched bearer services. The current alternatives for data communication in ISDN are summarized in Fig. 2.4. The circuit mode in ISDN is described in Section 2.2.1. This service provides a digital bit pipe where the user channels are presented to the users at the physical layer. A channel occupies an integer number of time slots (TSs) and the same time slot positions in every frame. A circuit is formed by the allocation of TSs to a channel in consecutive TDM frames available at the interface and connecting this to the destination by similar allocation of TSs in TDM trunks *en route*. The type of protocols to be used is left to the users. This provides a challenge to users as to how best to utilize this flexibility. The good BER of ISDNs means that light-weight protocols could be used at the data link and the network layers, hence reducing communications overhead. The packet modes in ISDN are described in the following section.

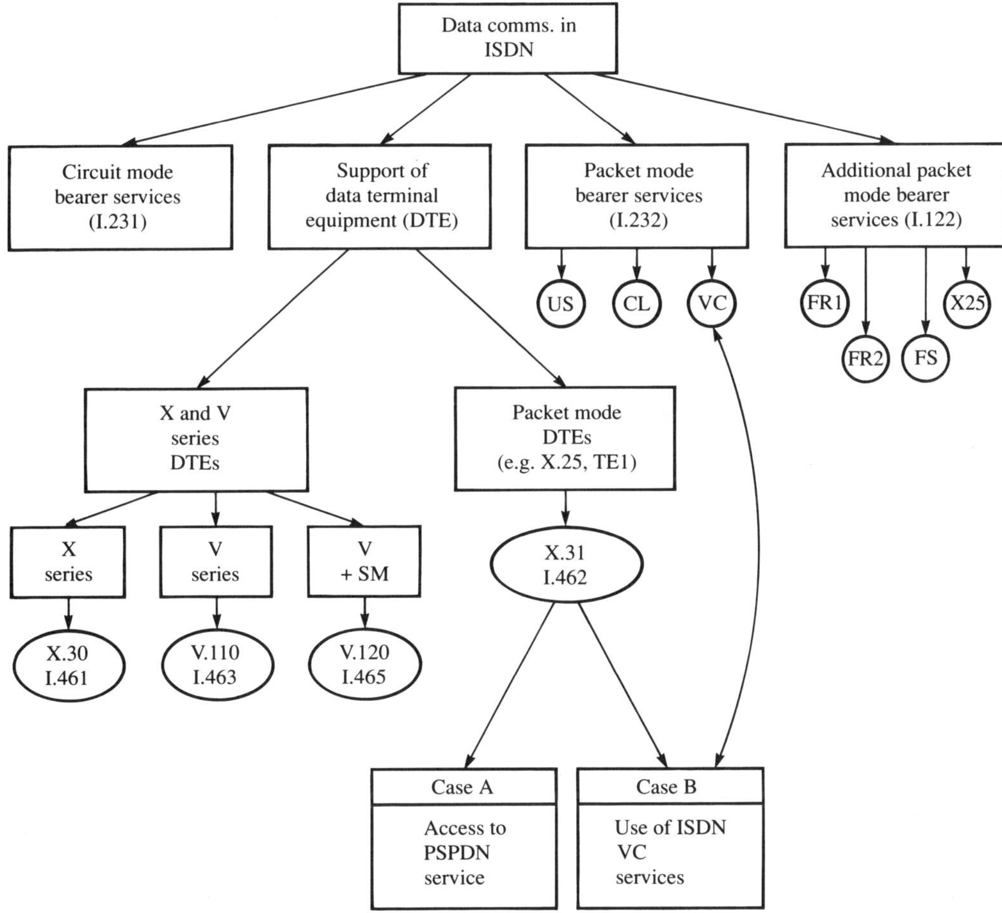

US: user signalling; CL: connectionless; VC: virtual circuit
FR 1/2: frame relay 1/2; FS: frame switching; SM: statistical multiplexing

Figure 2.4 Summary of data communication alternatives in ISDN

2.4 PACKET MODES IN ISDN

As defined by the CCITT Red Book (CCITT 1984), the ISDN is principally a circuit-switched network which can be used for packet network interworking. The definition of a framework for providing additional packet mode services (I.122) in the Blue Book (CCITT 1988a) has improved the data communications aspects of ISDNs. CCITT Rec. X.31 describes how an ISDN may offer access to an existing PSPDN from X.25 terminals connected to the ISDN via a terminal adaptor. It also describes how an X.25-like service may be supported wholly within an ISDN to offer a virtual circuit bearer service; however, this requires a two-stage call set-up and as such is awkward. The recent developments and draft proposals indicate that a move is being made to incorporate packet network access in a much more tightly coupled manner than access to some distinct packet-switching function or packet handler as was previously defined (Atkins and Cooper 1989). This will be a closer step to the provision of packet mode services wholly within the ISDN, which will be achieved by the expansion of the D-channel layer 3 protocols to cover packet-switching cases.

The CCITT Study Group XVIII has agreed on an evolutionary framework within which packet mode services would be considered. Accordingly, four phases have been described. Phase 1 incorporates enhancement of X.31 services described in the Red Book. Phase 2 includes the extension to X.31 service based on layer 2 multiplexing in the B-channel for an access arrangement to a packet handler and the use of X.25 PLP at layer 3. Phase 3 covers the additional packet mode services[2] based on full out-band call control and layer 2 multiplexing (I.441—LAP-D) on the B-channel. Phase 4 covers broadband services, including asynchronous transfer mode (ATM) and synchronous transfer mode (STM) services. Although Rec. X.31 (1984) describes a possible packet network internal to ISDN, it concentrates on the provision of interworking to existing PSPDNs. In line with the Blue Book, I shall describe the packet network interworking (or access to PSPDN services) and the VC bearer services separately. In addition to revisions to the X.31, foundations of phases 3 and 4 have been laid during the 1984–8 study period, and these too are described below.

2.4.1 Packet network interworking

Recommendation I.462/X.31 (Red Book) describes two different access scenarios to an X.25-based packet network (PSPDN): minimum and maximum integration scenarios. In the Blue Book (CCITT 1988a) these are classified as 'Access to PSPDN Services' (case A) and 'ISDN Virtual Circuit Service' (case B) respectively. In the minimum integration scenario, PSPDN access is via access units (AUs), which are situated within the PSPDN. In this case, the ISDN provides transparent, or rate-adapted connection, for an X.25 terminal to/from a PSPDN port (AU). Therefore, as such it offers access to non-ISDN packet mode services. This access could be semi-permanent or switched. For switched access, a two-stage call set-up procedure is needed. In the first stage, ISDN D-channel signalling procedures (I.451) are used to establish a connection to/from the AU. Then X.25 call set-up procedures are used over the circuit-switched connection to complete an X.25 call over the PSPDN. The AU will be allocated an ISDN address. The user data access will necessarily be over the B-channels.

[2] To be renamed frame mode services (FMS).

2.4.2 ISDN virtual circuit bearer service

This service is defined in Rec. I.232 (Blue Book) (CCITT 1988a). It was named the 'maximum integration scenario' in Rec. I.462/X.31 (Red Book). This provides an X.25-like service within an ISDN from the user's point of view. The Blue Book version also allows for access to a PSPDN supporting ISDN numbering. Packet terminals using the ISDN numbering plan will have their protocols terminated by an ISDN packet handler (PH) within the ISDN rather than the PSPDN access unit. Access to the PH may be via the B- or D-channel.

In access via the B-channel, a two-stage call set-up is needed. The user will not be aware of this because of the initial 'packet transfer mode' requested using the I.451 signalling protocol. The ISDN will then establish a circuit-switched connection to a PH. For access using the D-channel, LAP-D frames identifying packet data are interleaved with the signalling information. The distinction is done using a service access point identifier (SAPI) within the frame header. The network will route packet frames to/from the PH. Signalling frames are given priority over packet frames. An X.25 terminal uses a terminal adaptor (TA) to interface to the network termination. Throughput of data over D-channel is limited by the 16 kbps data rate, and additional delays may be incurred in the TA. However, it can be used in parallel with the circuit-switched B-channels.

2.4.3 Additional packet modes

The signalling scheme defined in the Red Book for the D-channel is for controlling circuit-switched user traffic channels and non-connection-related applications. The new Rec. I.122 (Blue Book) (CCITT 1988a) aims to use the enriched call control technique in controlling packet-switched virtual circuits. Recommendation I.122 describes only the architectural framework and service descriptions of the associated bearer services. It will apply to data communications at bit rates up to 2 Mbps. The reasoning behind the additional packet modes include the following: the provision of packet mode bearer services, which would be accessible using clearly separate control and user planes employing the same signalling protocols as other ISDN bearer services; and fast data transfer (with reduced delay and high throughput), by avoidance of full three-layer termination at each node in the network.

The method involves the use of outslot signalling with single-step call establishment in accordance with I.451, rather than the X.25 PLP call control procedure. Also, the multiplexing and switching of virtual connections takes place in layer 2 (LAP-D over B-channels) rather than the X.25 PLP. In the control plane, information elements specific to packet switching have to be added to existing I.451 signalling messages. In the user plane, the LAP-D functionality which will be terminated in every exchange may be restricted to a core subset including the frame delimiting, alignment and transparency; frame multiplexing and demultiplexing using the address field; and detection of transmission errors. These core functions result in a low-cost service, independent frame-relaying mechanism on a link-by-link basis which will result in low overheads and could be used in various applications.

Recommendation I.122 (CCITT 1988a) specifies four potential bearer services: frame relaying types 1 and 2, frame switching, and X.25-based additional packet mode.

- Frame relaying involves establishing a transparent network path for a call at call set-up time and then conveying subsequent frames along that path checking only the validity of frame format. Frame order is preserved, but frame acknowledgement within the network is not undertaken. Additional network functions to monitor and enforce throughput and achieve congestion control as well as frame error detection and QoS parameters are left for future study. The difference between the types of frame relaying is that type 1 implements only the *core* functions of Rec. I.441 (layer 2), whereas type 2 implements all the extended I.441 functions within the user terminals. In both cases, the network implements only the core layer 2 functions.

- Frame switching is similar except that the network passes on frames which correctly observe the layer 2 protocols. Additionally, it provides frame acknowledgement, error detection and recovery, and provides flow control. It necessitates the full implementation of layer 2 protocol (I.441 with extensions if necessary) both in the user terminal and in the network.
- X.25-based service adds to this the network support for X.25 PLP data transfer part.

Many important issues in the provision of additional packet modes still remain for future studies (see Atkins and Cooper 1989), including throughput enforcement, flow and congestion control (due to lack of layer 3 congestion and flow control mechanisms at layer 2), signalling enhancements to I.451, application categories (to determine type of bearer service), and charging (frame types need to be distinguished to enable charging).

2.5 PROTOCOL CONSIDERATIONS AND STANDARDS

With the introduction of distinct user and control planes in the ISDN reference model, functionally separate interactions associated with the control and signalling functions and transferring data are divorced, leading to separate and parallel developments of their functionalities and protocols. Within the CCITT Red and Blue Books, the protocols for the D-channel have been defined for the layers 1–3. For the B-channel, in the case of circuit mode services only layer 1 has been defined, while for the packet mode DTEs layers 1–3 based on the X.25 model are proposed. Additional packet modes incorporate layers 1–2 and 1–3 depending on the type of service, with LAP-D in layer 2 and X.25 PLP in layer 3 (when used). In the circuit-switched ISDN case, the user is left free to decide which particular protocol stacks to use on the B-channels for data communications over the links established. The D-channel signalling protocol is message/frame-structured and layered according to the OSI seven-layer model. Basic and primary-rate interface layer 1 specifications are necessarily different, but are common for layers 2 and 3. Table 2.2 shows protocol stacks on the D-channel for both BRA and PRA.

Table 2.2 Protocol stacks for the D-channel

Protocol	Signalling protocols	
layer	BRA	PRA
Layer 3	I.450 / I.451	I.450/1 (Q.930/1)
Layer 2	I.440 / I.441	I.440/1 (Q.920/1)
Layer 1	I.430	I.431

Recommendations I.450/1 (alias Q.930/1) define the network layer features and protocol. This has 'broad' similarities to the X.25 PLP functionality (e.g. frame formats), but is much more complex. It has a sophisticated message set in order to achieve call control for different types of services (bearer and teleservices, e.g. circuit- and packet-switched; additional packet mode services) and telephony functions. The Red Book focused on elementary call control, while the Blue Book enhanced this recommendation. However, the protocol is easily expandable to support more complex calls, standardized supplementary services and network specific services. Furthermore, these signalling procedures may also be used within private networking arrangements (e.g. between two PABXs).

Recommendations I.440/1 (Q.920/1) define the data link layer procedures for the ISDN D-channel (LAP-D). This protocol extends the X.25 LAP-B protocol. The main difference is that it allows multiple data links by virtue of an extended address field (two octets). It also uses several additional HDLC commands and responses.

Recommendations I.430 and I.431 define the physical characteristics of the basic and primary-rate user–network interfaces, respectively. The basic rate interface supports point-to-point as well as passive bus (S-bus) arrangements (including collision detection and contention resolution) and it has a four-wire 'S' interface. The primary-rate interface specifies coaxial wires for receive and transmit paths, and only the point-to-point configuration is supported according to CCITT Rec. G.703 (CCITT 1988a).

2.6 BROADBAND ISDN

Recommendation I.121 (Blue Book) on 'Broadband Aspects of ISDN' describes the basic concepts, service aspects, user–network interfaces and architecture models of B-ISDN. Asynchronous transfer mode (ATM) is chosen as the target transfer mode for B-ISDN, while synchronous transfer mode (STM) will also be supported in the interim. Unlike the narrowband ISDN (based on 64 kbps), it employs optical fibres for customer access. It proposes that all access to broadband and 64 kbps-based services will be via a single user–network interface S_B. Control of all services will be by means of CCS in line with the existing ISDN principles.

2.6.1 B-ISDN services

Two broad service categories are identified; interactive and distribution services. The interactive services are subdivided into conversational services (bidirectional end-to-end information transfer), messaging services (user-to-user communication via store and forward or mailbox type functions) and retrieval services (interactive videotex, video, text and graphics). The distribution services are subdivided into services with or without user individual presentation control.

2.6.2 B-ISDN user–network interfaces

Two types of user–network interfaces are proposed:

1. 150 Mbps interface: mainly for interactive services; symmetrical, and providing an H_4 channel plus some B- and D-channel capacity.

2. 600 Mbps interface: for interactive and distributive services; may be asymmetrical.

It is expected that the initial emphasis will be on the 150 Mbps interface as this will be preferred for fast data transfer because of its reduced transmission costs compared with the higher bit rate transmissions.

The following additional interface structures are proposed (though these channel rates will not appear physically at the S_B–T_B interface owing to the ATM principle of variable bit rates):

- H_{21} and H_{22} channels (32.768 and 43–45 Mbps respectively)
- H_4 channel (132–138.240 Mbps)

In addition to these, the S_B–T_B interface will have B, H_0, H_{11} and H_{12} narrowband channels converted to ATM and separate virtual signalling channels. The narrowband S–T interface can be provided by carrying the corresponding physical channels as virtual channels over the broadband access (Frantzen 1987).

2.6.3 Asynchronous transfer mode (ATM)

ATM is a new multiplexing and switching concept which is capable of handling both the bursty and stream type of traffic. Its most important feature is the flexible bandwidth allocation scheme, which decouples the interface from the network design. ATM is based on virtual channels (CO), with varying bit rates determined by service needs, and employs packet-switching technology. Fixed-size cells consisting of a header and an information field are used to transport data, reducing complexity and variation in delay. The basic ISDN protocol architecture regarding the separation of user, control and management planes is preserved. It is chosen by both the CCITT and CEPT as the target solution for B-ISDN.

The parties involved in its definition include CEPT (NA5), CCITT (Study Group XVIII) and the United States (T1/S1). On the technical side, some problems exist regarding the provision of time transparency (e.g. smoothing of propagation delay variations) and semantic transparency (e.g. error processing) in ATM networks. On the standardization side, so far the cell structures and ATM protocol layers have been defined in the Blue Book (CCITT 1988a). The cell and header sizes, call delineation, synchronization, flow control, congestion handling and adaptation between ATM and non-ATM parts of ISDN require further study.

2.7 CURRENT STATUS AND FUTURE TRENDS

2.7.1 Current status

The Recommendations developed during the CCITT study period 1980–4 (Red Book) concentrated on general principles and on a definition of ISDN services as well as the definition of D-channel protocol specifications. During the CCITT study period 1984–8 (Blue Book) much effort was put into further extensions. These included the extension of D-channel signalling protocols, the development of further network standards for ISDNs (numbering and addressing, network and terminal identification, network interworking and routing), the ISDN protocol reference model and consolidation with the OSI-RM, definition of additional packet modes, and the work on the definition of the B-ISDN principles.

One of the work areas that has progressed during the second study period has been the Q.931 expansion for layer 3 specification. This includes the essential features, procedures and messages required for call control (CS, user–user, PS) in the D-channel. However, some procedural details have not yet been specified and have been left for further study (e.g. transport or other message-based information flows, and the alignment of the functions and protocol with those of OSI NL). Further developments have been achieved in the CCITT Study Group regarding the subaddressing and NSAP addressing issues. These appear in the Blue Book as Rec. I.334 and propose, among other things, the conveyance of NSAPs in the Q.931 subaddress field. ISDN routing principles are detailed in Rec. I.335. Results of the work on the Q.931 and SSN.7 alignment (ISUP–ISDN User Part) as well as new interworking arrangements with existing networks also appear in the Blue Book (CCITT 1988a).

During the period 1984–8, the Red Book (CCITT 1984) provided enough impetus and reasonably firm standards for integrated circuit manufacturers to start work on new ISDN VLSI chips. Several manufacturers have developed and introduced ISDN chip families, board-level products and test instruments for the basic and primary-rate interface structures. It can be expected that Blue Book (CCITT 1988a) recommendations will further proliferate product and service implementations for the narrowband ISDN as these recommendations become even more stable.

Further enhancements and clarity were brought to the existing Rec. I.462/X.31 in the Blue Book. This is expected to increase the development of terminal adaptors for X.31 access to ISDNs as well as carrier provisions for such a service. However, it seems that X.31 (case B—i.e. the maximum integration) is the only recommendation on which the ISDN packet services can now be based. It is also to be used as the basis of early European functional standards for ISDN. There seems to be little interest in the UK for the provision of a service based on X.31 (case B).

Recommendation I.122, providing a framework for the additional packet modes (APMBS), has appeared in the Blue Book. Recommendation I.232 for packet mode bearer services, details three further bearer service categories: virtual call and PVC, connectionless, and user signalling. (Deletion of the last two services has been proposed to the CCITT Study Group XVIII.) These will add to the existing X.31 option for data transmissions within ISDNs and should prove that the common channel signalling on the D-channel can be used for all types of connections within the ISDNs. Its implications for hardware and software need to be evaluated. Several studies have so far been carried out for the use of APMBS in LAN interconnection. These studies include a Temporary Document from British Telecom (TD17 : BT to CEPT/NA1-WP3 Report) and the IEEE IVD–LAN Interface Working Group (IEEE 1988a).

Recommendation I.121 on the broadband aspects of ISDN, which appears in the Blue Book, is intended to provide a framework for the service definition, but there still is a lot of confusion regarding the technical implementation details (Minzer 1989) and B-ISDN applications despite the European Commission's aim to hold field trials during the 1991–5 period. However, we cannot expect to see much proliferation in the equipment or services available to customers before 1995. During the third CCITT study period (1988–92) the consolidation of the technical aspects of ATM and B-ISDN are expected. Without these, it is difficult to see manufacturers committing themselves to interim standards.

The CCITT Study Group XVIII prepared a new list of issues to be pursued during the 1989–92 period. These included the ATM; performance aspects; the interworking of ISDN and PSTN, PSPDN, CSPDN, ISDN; and private networks including LANs and user–network interfaces; layer 1 characteristics of B-ISDN; layer 1 characteristics of 64 kbps ISDN including the updating/completion of Rec. I.430/431.

The earliest option available for data transmission over (narrowband) ISDNs in Europe will be the circuit-switched access (as this is 'inherent' to the system). The second option will be packet-switched (X.31-based) access (though this will depend on the carriers' implementation policies). It seems that, for the X.25-based applications, the X.31-based implementations will be the norm for the immediate future. This protocol is now quite mature, although further evolution may be expected. The frame relaying/switching introduced in the Blue Book was still of a framework nature. However, some early implementations are available in the United States as frame-relaying 'islands'. Further clarity is needed before field trials of such a service can be held. Earliest B-ISDN field trials will be in the 1991–5 period. Again, this will depend on the agreements on the technical issues during the 1988–92 period.

2.7.2 Future trends

Customer applications will probably determine and justify the selection of bearer services, in which case some of the existing Recommendations may or may not be implemented by all carriers. The degree of acceptance of Rec. I.462/X.31 will depend on the success of the additional packet modes, since the frame-relaying/switching may be preferred for some applications and the X.25-based approach for others where a common call control procedure will be advantageous to the X.31 methods. The X.31 methods provide an early solution to the packet modes in ISDNs, while the use of X.25 over highly reliable (good BER) ISDN lines seems wasteful. Hence light-weight protocols for data transfers are needed. Additional packet modes of the Blue Book will provide such an alternative. However, it is still by no means certain that all the bearer services proposed in the additional packet modes will mature to be implemented. X.31 is still evolving, and ISDNs capable of supporting X.31 are currently being implemented.

The future of circuit-switched services within the ISDNs seems to be guaranteed (essential for telephony, fax and video), and they may be incorporated into the ATM-based ISDNs as well. However, it is interesting to note that the connectionless packet mode of working within the ISDN is not available. The alternative seems to be a move towards an all-CO mode of communications, both at the local networks and at the public networks, or the provision of a CL mode of operation on top of circuit-switched or semi-permanent lines or within some frame-relaying/switching or X.25-based 'tunnel'. Another solution may be the full CL and CO interworking arrangement, as indicated with the recent work by the ISO (1989).

Another issue is the recent and future developments regarding the ATM and B-ISDN. Co-ordinated B-ISDN field trials in Europe are expected during the 1991–5 period; an earlier start apparently could not be made because of the difficulty of deciding on the right multiplexing technology to be used. No one yet has a clear idea of how broadband will be used in the near future; the only existing demand is for high-speed data (and to some extent for video). One of the main problems the B-ISDN activists have been facing is finding real applications for B-ISDNs (e.g. motion video, medical applications, etc.). The ATM is a statistical technique and as such its capability to provide circuit-switched (CS) services are questioned. It is expected that it will eventually be able to replace CS services, even for voice. However, the costs involved are prohibitive for now. Another problem of the ATM technique is not so much the speed, but the handling of a large number of basic channels like the 64 kbps narrowband ISDN channels.

Value-added services (VASs) depend on the provision of higher-layer functions associated with the OSI layers 4–7 and are provided on top of basic services by value added carriers. The VAS service modules or servers can be accessed from a number of different networks, including the ISDNs. These services will include database access, store-and-forward communications, communications support services and compatibility services. The enhanced signalling capabilities of an ISDN is seen as a useful method for allowing signalling messages to be exchanged between a VAS user and server in parallel with, or in the absence of, an established network connection. It is thought that ISDNs will provide an optimum framework for VAS. Hence, widespread implementations of VASs are expected in the ISDNs in future.

2.8 RELEVANT FEATURES OF ISDN IN LAN–ISDN INTERCONNECTION

In the design of data-oriented LAN–ISDN relays, the following features of the ISDN are most relevant:

- Multiple channel access structures
- Common channel signalling (CCS) using the D-channel
- Switched and non-switched capabilities
- Circuit and packet-switching capabilities

Multiple channels within one interface, which are user-controlled with regards to the way they are used and the type of ISDN services selected, are available in the ISDN era through the use of a CCS system. This is a completely new development in the field of data communications, where most networks still use leased lines for interconnection. Packet data is by nature bursty and the bandwidth requirements of data networks connected to ISDNs may vary during a given time duration. Hence one could conclude that the channels at the UNI could best be utilized by dynamically controlling their total bandwidth. This leads us to the idea of superchannels and dynamic channel (bandwidth) management, issues that are studied in Chapters 4 through 10.

Naturally, the integration of services within the ISDN also has implications for the integrated services LANs (ISLANs) and integrated services PABXs. However, these issues are not further discussed in this book since here we are focusing on the data applications across the ISDN itself.

2.9 SUPERCHANNELS IN ISDN

In this book, channels of arbitrary bandwidth size (in units of 64 kbps) are termed 'superchannels'. Current ISDN recommendations do not specify such a facility, although channels at predetermined higher bandwidth values are defined. However, there is a need in many applications to have channel capacities that are not limited by the basic channel types.

Hence formation of superchannels by suitable aggregation of basic channel types is desirable. However, identification of such channels on an end-to-end basis, as well as provision of dynamic variation of their allocated bandwidth, is not yet available within the CCITT recommendation. This issue is addressed in Chapter 6 and solutions are proposed.

2.10 TARIFFS

This issue has not been completely addressed so far by the parties involved. However, the early indications show that, for data communications using the circuit-switched ISDN, a tariff structure similar to that of telephony will be applied. This means that each call is charged according to the charge band applicable (depending on the time of day) and distance. Each unit is charged to the customer at the beginning of the charge period. In some countries the charge period (or the charge applied to each charge period) is not linear, the first charge period being shorter (or the first charge unit being higher), so that a type of 'connection cost' is incurred by the users. In the rest of this book we will assume a linear charge period.

2.11 SUMMARY

This chapter provides a background to the ISDN by first discussing its general principles and important technical features and then presenting a short summary of the CCITT Recommendations pertaining to the salient features of ISDNs. The section on the network performance objectives for call processing delays for circuit-switched connections indicates the type of performances to be expected from ISDNs. This will be useful later in relating to the simulation models developed in Chapter 8. The ISDN Protocol Reference Model includes the logical user and control planes and proposes a plane management function for co-ordinating their activities. It is with this background that a channel management architecture is proposed in Chapter 5. The relevant features of ISDN for LAN–ISDN–LAN interconnections are also presented, forming the basis for our discussions in Chapters 3, 4 and 5. Here, it is noted that bandwidth management is not included in the current ISDN UNI Protocol for Management (Q.940). The short description of superchannels is intended to form the background to Chapter 6, where superchannel issues are discussed in detail.

CHAPTER

THREE

INTERCONNECTING LANs AND ISDN

Developments in the local and wide area carrier technologies, and the need for new applications with large bandwidth requirements, running on very fast processors, mean that the methods of interconnection of these devices will be the crucial issue in their collective performance. Today, in the local area, the workstation/server model with fast local area network (LAN) interconnection has become the norm in data communications and local networking. In the wider area, private and public networks with ever increasing capabilities are providing better facilities for the interconnection of local networks, servers and workstations.

In data communications, the interconnection of LANs using ISDN has become an issue of great importance because of the ISDN capabilities described in Chapter 2. The ISDN, apart from integrating services and providing common user interfaces, will facilitate fast switching via end-to-end digital connectivity as well as multiple channels at various transmission capacities. The use of common channel signalling (CCS) will mean that control of these channels can be made in an outband mode concurrent with the user data transmission over the user channels. To date, little has been reported about how these facilities may be used in LAN–LAN interconnections using the ISDN.[1] Hence our interest in LAN–ISDN–LAN interconnection arises from two ambitions: to investigate the issues involved in the LAN–ISDN interconnection, and to study the channel management issue at the boundary of the packet switching–circuit switching network interconnection. The LAN–ISDN interconnection will be a common source of PS–CS network traffic.

[1] In this book, a distinction is made between the LAN–ISDN and LAN–ISDN–LAN interconnection. In the broadest sense, the former can be assumed to support the integrated services facilities of the ISDN. However, since most existing LANs support data-only services, and since this book is concerned solely with data communications, it will be assumed to refer to data applications only unless indicated otherwise. In LAN–ISDN interconnection, it is further assumed that remote ISDN workstations can access the LAN hosts and services transparently. In contrast, the latter implies that the ISDN is transparent to the LAN applications and supports only data communications. This means that the two end LANs are more tightly coupled, and the hosts and workstations attached to the ISDN may or may not be able to access the LAN services. (See Sections 3.2 and 3.3 for further discussions.)

Section 3.1 presents a short overview of general issues involved in packet network interconnection as well as issues involved in the interconnection of LANs and packet networks. Section 3.2 studies the type of ISDN services required for different levels of interworking with the ISDN. As already stated, the main area of interest in this chapter is the interconnection of LANs using ISDN; various alternatives to achieving this objective are studied in Section 3.3. Here, because of its earlier availability and ubiquity, I concentrate on the circuit-switched ISDN while summarizing other options. The following section looks in particular at LAN–LAN interconnections using the primary-rate access[2] and highlights the usage of multiple channels at the ISDN user–network interface. This serves as a groundwork for the investigation of dynamic management of multiple channels in Chapter 4. Finally, a summary of the chapter is presented.

3.1 PACKET NETWORK INTERCONNECTION ISSUES

For a multitude of reasons which have already been discussed above, computer networks need to be interconnected (see Fig. 3.1). The common objective of all interconnection methods is to allow all subscribers a transparent means of accessing a host or service on any of the interconnected networks.

To achieve this objective, data produced at a source in one network must be able to be delivered and correctly interpreted at the destination(s) in another network. This necessitates the provision of inter-process communication across the network boundaries (Cerf and Kirstein 1978). In order to achieve this, some sort of 'commonality' is required across the interconnected networks. This can be achieved by two general approaches:

1. Translation of one protocol into another, when different protocols operate on separate networks.
2. Using protocols common among the communicating parties.

Also, a strategy combining the two approaches may be employed (e.g. using some common protocol at a given layer and translating protocols at one or more layers to the interworking protocols). In order to provide a common model for data communications, ISO has developed the Reference Model for Open Systems Interconnection (OSI-RM) (CCITT 1988a; ISO 1984). This model provides the basis for development of common protocols as well as for interworking between heterogeneous environments (Burg and Iorio 1989).

3.1.1 Network interconnection approaches

From the point of view of the desired functionality, three general approaches to network interconnection can be described (Schneidewind 1983):

1. *Network access* Achieves physical access and protocol compatibility in the lower layers of the OSI model (e.g. physical, data link and network layers of X.25). This method is used when the networks (e.g. LANs and the long-distance network) already exist with their own established protocol structures.

[2] It also applies to the basic-rate access.

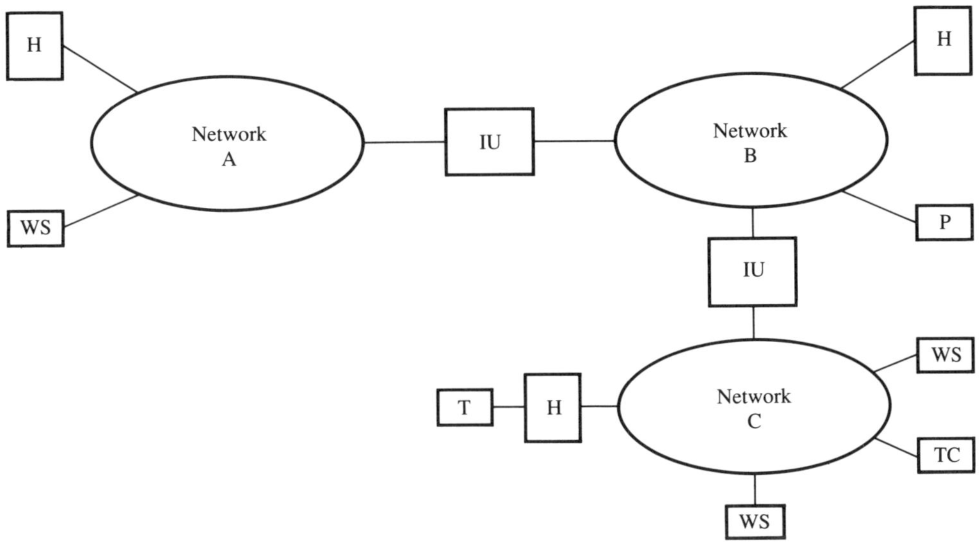

IU: interconnecting unit (e.g. gateway); H: host/server
WS: workstation; T: terminal; TC: terminal concentrator; P: pad

Figure 3.1 Multiple network interconnection

2. *Network services* Obtains the use of specific types of services (e.g. virtual circuit service) to satisfy user needs. This method is used for resource sharing or value added networks. It stresses obtaining network services for the user (e.g. provision of remote interactive processing involving session control and sequenced error-free message delivery).
3. *Protocol functions* Matches the number and types of protocols on each type of network. Protocol conversion is necessary for this method, since different sets of protocol functions are assumed at each end.

Two similar terms are usually used in describing the interaction between two or more networks:

- Network interconnection
- Interworking (or internetworking)

In this book we define the network interconnection issues as those relating to problems and general methods of interconnecting different types of networks without necessarily achieving full interworking. Interworking issues are defined as those issues relating to the provision of services and facilities available on one network to the users on another network or to single users accessing that network using isolated terminals or workstations.

The network access approach falls into the interconnection classification, while the network services falls into the interworking classification described above. In a layered communications architecture, the protocols used in each layer provide a service to the layer above. Hence the protocol functions approach falls somewhere between the the two. In this book we are interested in network access as far as the LAN–ISDN interconnection is concerned. For the LAN–ISDN–LAN interconnection, however, we are interested in the LAN–LAN interworking. Hence the use of ISDN to interconnect LANs and isolated individual hosts/workstations is investigated from this viewpoint in Section 3.3.

3.1.2 Technical issues in interconnection

The following issues need to be resolved for a coherent network interconnection:

- Level of interconnection
- Naming, addressing and routing
- Flow and congestion control
- Access control (security)
- Common services (e.g. internet services)

Other issues such as buffering, fragmentation and re-assembly, multiplexing and error control, must also be considered (see Postel 1980; Sunshine 1990).

Different networks can be interconnected by 'mediating' systems that are (generically) called *gateways*, corresponding to *relays* in the OSI terminology (see Section 3.1.4).[3] The fundamental role of a gateway is to terminate the internal protocols of each network to which it is attached, while at the same time providing a 'ground' across which data from one network can pass into another. Gateways can be used in the following strategies (Cerf and Kirstein 1978):

- Packet-level interconnection (common subnet technology)
- Common network access interface (datagram and virtual circuit)
- General host gateways
- Protocol translation gateways

In general, when different protocols operate on separate networks, two approaches can be achieved for interconnection:

1. *Media conversion* A media conversion gateway bridges the gap between differing data link and physical layer protocols; messages from one network are read by unwrapping their network packaging, the necessary routing is computed, and messages are sent to another network by wrapping them into that network's packaging. Examples of these are repeaters and bridges operating at the physical and data link layers, respectively (see also Section 3.1.4).

[3] We note that a more general term for a mediating or intermediary system is *interconnecting unit* (IU) or *interworking unit* (IWU). There is a lot of confusion in the use of these terminologies. Another classification often used for different relay types (and their commonly used non-OSI terminology), which is adopted in the rest of this book, is (Perlman *et al.* 1988):

(i) Physical layer relay: *repeater*
(ii) Data link layer relay: *bridge* (Seifert 1988; Tanenbaum 1989)
(iii) Network layer relay: *router* (Seifert 1988; Tanenbaum 1989)
(iv) Higher layer relays (layer ≥ 4): *gateway*

We note that the generic term 'gateway' is also used widely in referring to the data link and above-layer relays.

2. *Protocol conversion* A protocol conversion gateway bridges the gap between differing network and higher-layer protocols; messages received from one network are replaced by different messages with the same protocol semantics and sent into another network. Examples of these are network or transport layer relays.[4]

In the OSI approach, transport gateways are an anathema, since they are not supposed to be needed because network interconnection is expected to be at layer 3 or below. Gateways are not discussed in any detail in the rest of this book, since the OSI model of interconnection described below is adopted. Furthermore, as the main focus of the book is on the dynamic channel and bandwidth management, as described in Chapter 4, the main interconnection issues considered are the level of interconnection and those arising from the LAN–ISDN interconnection. These are detailed in Sections 3.1.3, 3.2 and 3.3.

3.1.3 The OSI model for interconnection

The OSI-RM (CCITT 1988a; ISO 1984) refers to a seven-layer architecture. The top three layers (layers 5–7) are responsible for the processing of information, while the bottom four layers (layers 1–4) are responsible for moving information units. According to the definition in Section 3.1.2, a co-operation between heterogeneous networks at layers 1–4 provides interconnection, while the full 1–7 layer co-operation provides interworking between end systems and processes. It is the heterogeneity of networking technologies at both the local and wide areas that causes the interconnection problems, since information units must be transported across these networks securely. Interworking necessitates interconnection, but not vice versa. Heterogeneous networks can be interconnected by intermediaries. A *relay* described by the OSI-RM is such an intermediary. Its function is to facilitate communications between peers in the protocol hierarchy. A relay implements a set of procedures by which data from one system is forwarded to another. This is called a *relaying function*. A relay sharing a common protocol at layer n with other systems but not participating in a layer $n + 1$ protocol in implementing its relaying function is called a *layer n relay* in OSI terminology.

In the OSI-RM, layers 4 and above operate on an end-to-end basis. Furthermore, OSI principles state that interconnection of data networks (*subnetworks* in OSI terminology) must be achieved at layer 3, the network layer, since this layer provides global network addressing and provides routing and switching as well as relaying functions, among others. Network interconnection can also be achieved at layers 1 and 2; however, as these result in *extended subnetworks*, they do not contravene the OSI principles. These issues are further dealt with in Section 3.2.

The OSI-RM specifies that layer $n + 1$ makes use of the layer n service in performing its own layer service. This leads to the *n-layer service interface* and the *service relay* concept for network interconnection. This assumes that the services operating on either side of a relay are identical (e.g. connection mode network service—CONS). In this case, the relaying function

[4] Strictly speaking, two types of relays can be defined here: service relay, and protocol relay. For example, a transport service (T-service) relay would terminate two transport protocols and relay T-service primitives. A T-protocol relay does not terminate any protocols; it merely passes the semantics with a new syntax on to the next subnetwork.

consists of a simple mapping of indications and confirmations arriving through one service interface onto requests and responses sent through the other. Provision of different services on either side causes incompatibilities. In practice, the services offered by real subnetworks (e.g. LANs and X.25 PSPDNs) are rarely identical and usually they do not provide a true OSI network service (CCITT 1988d; ISO 1987a). A further complication is that the OSI Network Service (OSI-NS) can have two guises: the connection mode (CO) and connectionless (CL) network services. If these are operated on either side of a relay, a CO/CL interworking arrangement is needed (ISO 1989). This is often the case when LANs (usually CL) and WANs (usually CO) are interconnected.

According to the OSI principles, service incompatibilities on either side of a relay can be remedied by modifying the service on one or both sides such that a common service is achieved (Lenzini 1984). This aim is made easier to achieve since the OSI network layer functions are divided into three sublayers (Burg and Iorio 1989; ISO 1987c): two convergence protocols and a subnetworking access protocol (SNAcP).[5] The convergence protocols are referred to as the subnetwork independent and dependent convergence protocols, (SNICP and SNDCP)[5] respectively (see Figs. 3.2 and 3.3).

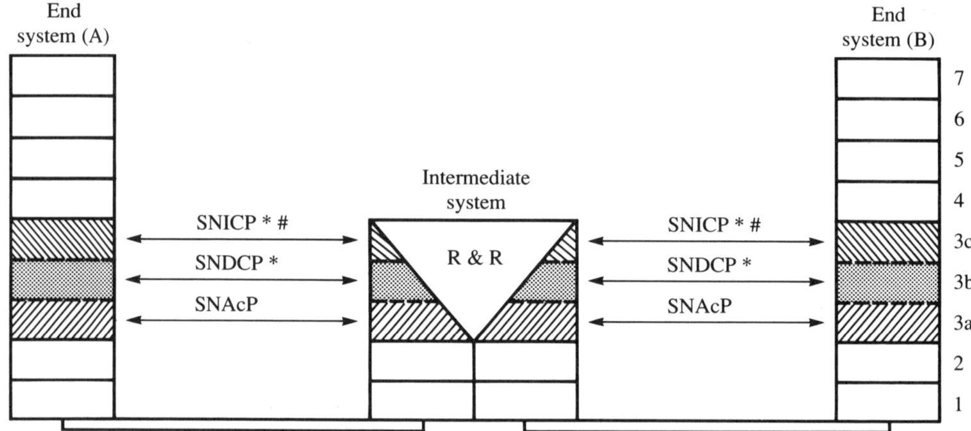

R & R: routing and relaying; SNICP: subnetwork independent convergence protocol
SNDCP: subnetwork dependent convergence protocol; SNAcP: subnetwork access protocol

* May not be needed in all instances.
# These protocols may be identical.

Figure 3.2 Partitioning of network layer

[5] An SNAcP handles the network layer functions for the specific subnetwork being used and could be different for other subnetworks. If a protocol acting in this role provides the OSI-NS, it fulfils all three roles; otherwise SNDCP and/or SNICP roles are needed. The SNDCP is designed to harmonize subnets that offer different services. If a protocol acting in the SNDCP role provides the OSI-NS, then it is deemed to be acting in both the SNDCP and SNICP roles; otherwise the SNICP role is needed to present the OSI-NS at its interface with the transport layer.

The roles assigned to different sublayers can be provided by a single protocol (e.g. the X.25 Packet Level Protocol (PLP)—as well as by a set of protocols. The role of the convergence protocols is to enhance or de-enhance an existing service to provide the OSI Network Service at the service interface to the transport layer. This extra layer of protocol needs to be added to every network node wishing to communicate with a NL relay configured in this fashion. A practical way to avoid the large number of modifications the above scheme implies is to implement relaying at the transport layer (see Section 3.1.4), where a commonality can be achieved (using either CO or CL network service modes). However, this compromises the end-to-end nature of the transport service and hence violates the OSI-RM.

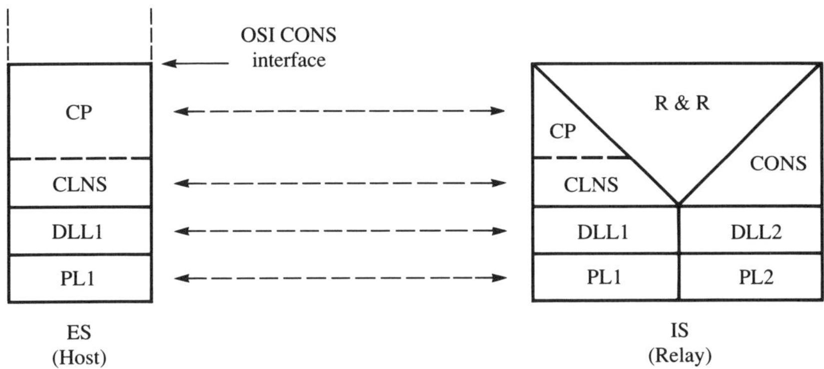

ES: end system; IS: intermediate system; R & R: routing and relaying
PL: physical layer; DLL: data link layer; CLNS: connectionless network service
CONS: connection mode network service; CP: convergence protocol

Figure 3.3 Use of a convergence protocol

ISO work provides a framework that can be used to interconnect both OSI and non-OSI networks to provide the OSI Network Service. Interconnection of networks whose SNAcPs provide the OSI-NS is straightforward because of the service compatibility across all participants. When interconnecting networks that include non-OSI networks, two approaches are available under the common service concept:

- Hop-by-hop harmonization
- Internetworking protocol

The hop-by-hop harmonization approach prescribes the 'harmonization' of the subnetwork service available on each non-OSI network to that of the OSI-NS. The harmonization functions reside in the SNICP sublayer and possibly, when (de-)enhancement is necessary, the SNDCP sublayer. Routing and relaying functions are employed to interconnect the harmonized subnetwork services to provide the OSI-NS end-to-end. With this approach, all end and intermediate systems attached to a given subnetwork have to support the same sublayer protocol(s) to enable communication between all end system in this subnetwork.

In the internetworking protocol approach, an internet protocol (operating in the SNICP role) runs over the interconnected series of networks, which may be of different types, providing the OSI-NS across these networks and intermediate systems. An internetworking protocol is based on a predefined set of capabilities over which it is to operate. If a network participating in the interconnection cannot provide a subnetwork service adequate for the capabilities required by the internetworking protocol, a sublayer SNDCP protocol has to be used to enhance its subnetwork service to the level of that required by the internetworking protocol. When this approach is used for network interconnection, the same internetworking protocol has to be used in all end-systems and relays to enable communication between these systems to take place.

Interestingly, the only internetworking protocol international standard is the connectionless network protocol (CLNP) (ISO 1988), resulting in the provision of the OSI connectionless network service (CLNS). Furthermore, for the provision of the OSI connection-oriented network service (CONS), the only methods applicable are the interconnection of subnetworks whose SNAcPs support the OSI-NS and the hop-by-hop harmonization. Present CCITT proposals support only the OSI CONS; hence it recognizes only the latter two methods (Burg and Iorio 1989; CCITT 1988f).

3.1.4 Interconnection at different layers

As mentioned earlier, interconnection of networks at layers other than the network layer is possible. Here, these are mentioned briefly. It is interesting to note that session or presentation layer relays have so far not been developed (see Sunshine 1990).

Data link layer relay The MAC bridge (IEEE 1988b) is a widely used example of such interconnection. It achieves relaying within layer 2 and interconnects LANs. Local and remote bridging techniques are available. More than two LANs can be connected via a single bridge. However, MAC layer bridges present problems in areas of address management, security broadcast control, resilience, load balancing and upgradability. Further problems arise in (global) addressing and protocol compatibility when LAN services are to be accessed by remote terminals or workstations.

Transport layer relay The CO/CL network service interworking without modifications at the hosts, can be established by using a transport layer relay. An ISO Technical Report (ISO 1989) proposes passive and active types of transport layer relays (TLRs). A TLR is called an interworking functional unit (IFU) and achieves interworking between CO and CL network services by relaying and/or conversion of protocol data units (PDUs) from one network type to another. This method assumes the use of a connection oriented transport service on both networks in order for them to be interconnected. However, these modes are outside the OSI architecture, and hence this solution is not an International Standard but a Technical Report.

Application layer relay Application layer relays are used for mapping largely incompatible (and often proprietary) protocol stacks, masking incompatibilities at lower layers. Examples are terminal and mail gateways. They may also be used for administrative reasons, e.g. to restrict inter-domain traffic to certain applications. Their major disadvantages are the need to develop new gateway modules for each application and set of protocol stacks, and slower performance.

3.2 INTERCONNECTING WITH ISDN

Two issues determine the level of interworking between an ISDN and an external user:

- Type of user interface (or user interface capabilities)
- ISDN services needed, ie. bearer and/or teleservices

The type of user interface determines the level of services that can be obtained from the ISDN. For example, a terminal or host supporting only traditional data interfaces, which are non-ISDN-compatible, needs to be connected through a terminal adaptor to the ISDN. This means that it cannot request the full range of services offered through an ISDN-compatible interface. Since teleservices require layers 4–7 compatibility with ISDN, these services are offered only to equipment supporting ISDN interfaces (compatible with S or T reference interworking). When interconnecting LANs and terminal or workstations to ISDN, four cases can be described:

1. Provision of all ISDN services to LAN users
2. Provision of LAN services to ISDN terminals/workstations
3. Provision of LAN services to non-ISDN terminals/workstations
4. LAN–LAN interworking across ISDN

The first case necessitates the provision of ISDN-compatible interfaces to the LAN users capable of interworking at the S reference point. This necessitates the support of the S-interface over the LAN and cannot be easily implemented over the existing data-only LANs. With such a functionality the LAN becomes more like an ISPBX. We shall call such a LAN an *ISDN-compatible* LAN (IcLAN). Also, a LAN may be modified to support integrated services, in which case it is called an integrated services LAN (ISLAN). However, an ISLAN does not necessarily need to support the S-interface.

The second case can easily be supported if the LAN is connected to the ISDN via a network layer relay. The third case necessitates the use of either the circuit or packet mode bearer services by both the LAN and the non-ISDN terminal/workstation for data transfer. Finally, the fourth case necessitates the access of ISDN switched or non-switched bearer services for interconnection and can be implemented to operate at the R, S or T reference point. Table 3.1 summarizes the ISDN services needed and or usable in each case.

In this book we are interested only in cases 2, 3 and 4. Here, the type of bearer service to be used and the layer of interconnection need to be clarified. These also have an effect on the protocol stacks to be used on the LAN/workstation as well as the ISDN. Furthermore, the way in which LAN is to be interconnected to the ISDN needs to be specified.

3.3 INTERCONNECTION OF LANs AND ISDN

The choice of reference points with regard to the usage scenarios described in the previous section are given in the following subsections. In all of these scenarios, we assume that the LAN is connected to the ISDN via a relay, as opposed to individual LAN hosts having direct access to the ISDN.

Table 3.1 Services in LAN/workstation–ISDN interconnection

Interconnection type[a]	ISDN services	
	Bearer service	Teleservices
T/WS–ISDN–Host/Server	√	–
T/WS–ISDN–LAN	√	–
LAN–ISDN–LAN	√	–
LAN–ISDN	√	–
IcLAN–ISDN	√	√
IWS–ISDN	√	√
IWS–ISDN–IcLAN	√	√

[a] T/WS: terminal/workstation; IWS: ISDN workstation; IcLAN: LAN modified/designed to support S-interface.

Interconnection at reference point R Figure 3.4 shows the LAN–ISDN interconnection at reference point R. The LAN–ISDN relay can have two types of interfaces: circuit or packet mode, hence acting as a circuit or packet mode DTE, respectively. In either case, a compatible terminal adaptor is used to interwork with the S-interface. Thus, usage cases 2, 3 and 4 can be supported with this configuration. Ordinarily TAs will support only one B- or H-channel at the ISDN side, thereby limiting the internet traffic that can be handled. However, intelligent TAs which can aggregate B- or H-channels in ISDN are also possible, and these could support larger-capacity connections.

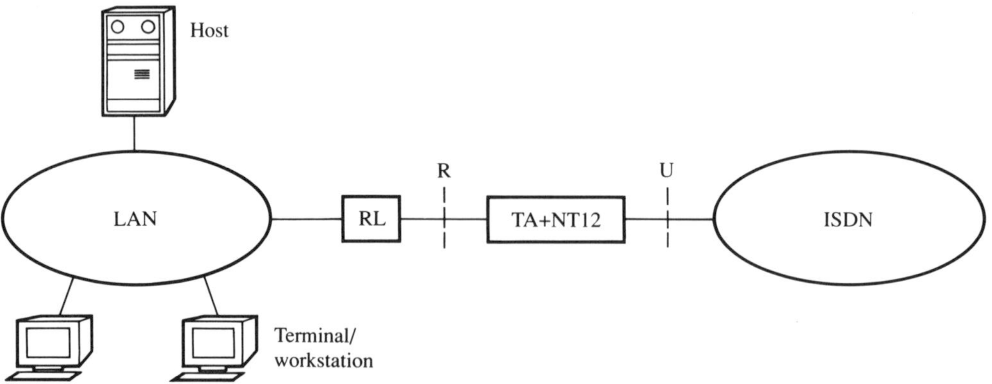

RL: relay; TA: terminal adaptor; NT12: network termination 1 and 2

Figure 3.4 LAN–ISDN interconnection at reference point R

Interconnection at reference point S This configuration, shown in Fig. 3.5, can be used to support usage cases 2, 3 and 4. The advantage of this configuration is that it provides direct access to the S-interface by the LAN–ISDN relay, which can use the flexibility of the control of that interface by ISDN signalling procedures. In this mode, LAN–relay combination acts as an intelligent terminal adaptor. This is the preferred interworking point for the purposes of this book.

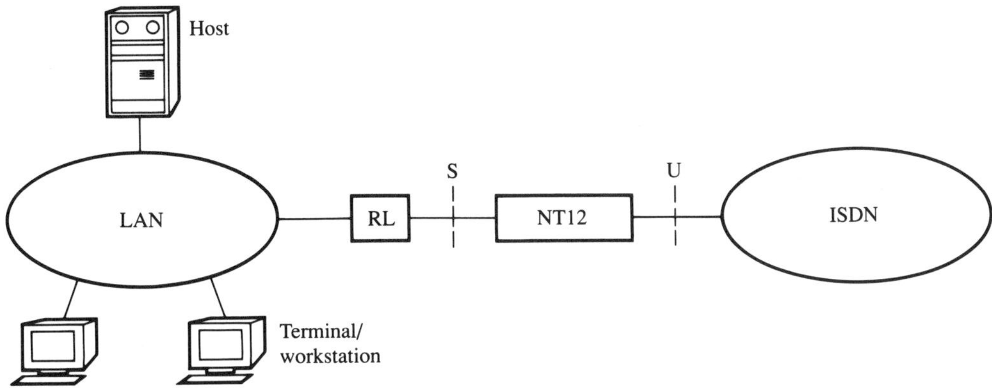

RL: relay; NT12: network termination 1 and 2

Figure 3.5 LAN–ISDN interconnection at reference point S

Interconnection at reference point T This is the only configuration where usage case 1 can be supported in addition to cases 2, 3 and 4. The LAN needs to be interconnected at reference point T to the ISDN (Fig. 3.6). The LAN acts as an NT2 and provides the multiplexing and distribution functions as well as the S-interface to LAN users. However, a LAN supporting the S-interface needs to be designed carefully in order to benefit from this configuration, as none of the current LANs provide this facility, nor can normal LAN hosts exploit it.

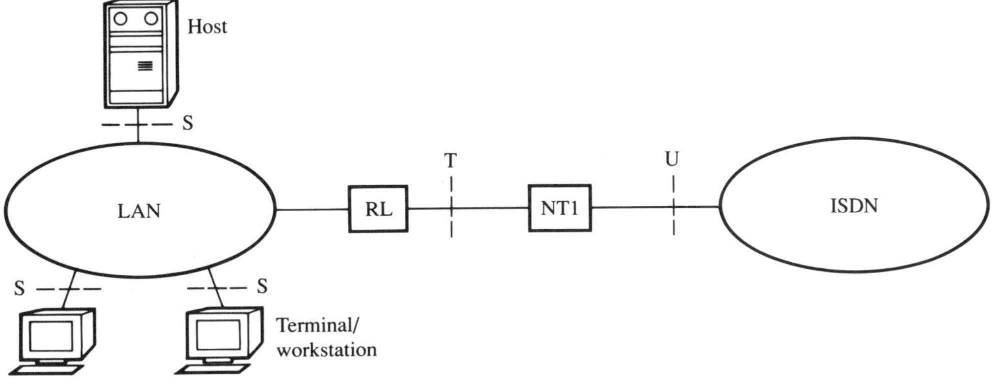

RL: relay; NT1: network termination 1

Figure 3.6 LAN–ISDN interconnection at reference point T

3.3.1 Using the ISDN bearer services

In Chapter 2, the data communications in ISDN were classified under the circuit- and packet-switched bearer services and the additional packet mode bearer services (APMBS). The packet-switched bearer service is currently available only using the X.25 protocol suite, and the APMBS is not expected to be available in Europe in the immediate future.

The frame-relaying type 1 (Lai 1989a, 1989b, 1989c) implies a core layer 2 service on the B-channels and is found to offer significant benefits for the interconnection of LANs by *remote bridges*. Indeed, its trade-off between functionality and speed appears to suit LAN–LAN interconnection better than either of the other alternatives, i.e. circuit-switched channels or X.25 packet switching. This is because the simple access interface, relatively high access speed and statistical multiplexing capability (in layer 2) of frame relaying is ideally suited to LAN interconnection (Lamont *et al.* 1989). The remote bridge solution using frame relaying suggests the encapsulation of LAN MAC layer frames in the LAP-D frames (Rec. I.441). High-speed operation is possible by transparent operation because of the non-termination of higher-layer protocols and the use of core layer 2 functions only (e.g. no frame acknowledgement). Owing to layer 2 multiplexing, high-speed access is more cost-effective, especially to multiple destinations. This facility is similar to the multiple virtual circuits in X.25 networks but at reduced protocol overhead and associated speed/cost penalty of layer 3 multiplexing.

However, because of the additional requirement of enabling isolated workstations to be able to access the LAN services and the earlier availability and ubiquity of the circuit-switched service, it is this ISDN bearer service that will be referred to in the rest of this book. Therefore, LAN interconnection using either the packet mode bearer service or the APM bearer service is not discussed further in this chapter.

A further requirement in LAN–LAN interconnections is adequate bandwidth provision. In cases where off-site traffic is light enough, a basic-rate access may suffice. However, when off-site traffic volumes are high and LAN applications best served by high-bandwidth links are to be used (e.g. fast file transfer, CAD, bit-map and multi-media document transfer), primary-rate access arrangements need to be considered. In the rest of this book, we shall assume the use of PRISDN.

Another issue in LAN–ISDN interconnection is the layer of interconnection. Following the earlier discussions, we shall assume that the interconnection is achieved at the network layer. Hence the network service is to be provided over the circuit-switched ISDN digital bit pipes. The protocol stacks that could be used over these bit pipes are discussed in Section 3.3.5.

3.3.2 Applications running on LANs

Although it is difficult to predict future communication needs and patterns of various user groups, all existing data communication applications could be supported over the ISDN. Furthermore, it is expected that remote terminal, file and database accesses, fast file transfer, distributed databases, CAD applications, electronic mail, multi-media document services and telematic services applications will be prominent in the immediate future. It is obvious that some of these applications will not be able to use the full 64 kbps channel capacity when offered, for example, a remote terminal access session. Several such sessions can therefore be multiplexed onto a single channel if they pass through the same pair of ISDN stations (access nodes). Also, interactive applications such as terminal access and file access need 'real-time' responses in order to be usable. Delay becomes more important than throughput in these cases.

3.3.3 Interconnection scenarios

The following LAN–ISDN–LAN interconnection scenarios can be visualized:

- *Scenario 1* A fixed number of LAN sites communicate with each other on a regular basis (e.g. branches of a corporation) and have regularly distributed traffic.
- *Scenario 2* Users on a large number of LANs want to communicate with each other on an irregular basis.

In scenario 1, a leased-line or semi-permanent line solution may be more cost-effective than a circuit-switched solution, as the advantages of ISDN will not be evident if the tariffing structure is not favourable to switched connections. However, the ISDN could provide added bandwidth on demand or act as an emergency backup facility. In scenario 2, a circuit-switched solution will be more cost-effective. Hence a judicious management of the circuit-switched connections under different tariff structures, user demand (traffic volume) and service/user priority constraints will bring about cost-effective utilization of the local ISDN bandwidth. It seems more likely that an advanced automated office environment will need the flexibility of scenario 2 for its applications.

3.3.4 The LAN–ISDN relay

The LAN–ISDN relay is a specialized relay, which must overcome several technical problems. It differs from most other relays in that:

- It is multichannel.
- It must manage many switched connections simultaneously.
- It has common channel signalling for outband control of circuits.
- It connects to a network (the ISDN) which imposes no restrictions on the protocols to be used—the B-channels appearing merely as bit pipes (assuming circuit-switched bearer service).

These properties present the following problems:

1. Intelligent management of multiple channels for switched connections; circuit set-up and removal actions as well as dynamic bandwidth control.
2. Efficient queue management for channels.
3. Conversion of inband data signalling (e.g. addressing, routing and other packet information) on the LAN to outband control signals for circuit-switched connection set-up on the ISDN. This is basically a way of protocol conversion from the single-plane format of the OSI-RM to the two-plane (user and control planes) format of the ISDN–PRM.
4. Choice of protocol stacks to be used over bit pipes.

In the local area, the workstation/server/LAN system model has become the most popular environment, and this must be interfaced to other LANs via the ISDN. A LAN–ISDN relay with a LAN interface on one side and a PRISDN interface on the other is needed. The capabilities required from this relay must reflect service requirements as well as the underlying communications requirements. The LAN–ISDN relay must be able to manage multiple channels which may be switched individually or in bundles, depending on user/network requirements. Procedures need to be defined to manage a variable number of B-channel connections depending on system status and traffic variations.

3.3.5 Protocols over the ISDN bit pipes

Another issue in the LAN–ISDN relay is the type of protocols to be supported. Both the CO and CL network services can be provided over the digital bit pipes provided by the ISDN circuit-switched bearer service. This leads to different protocol stacks in the LAN–ISDN network layer relay. For simplicity, the same type of network service is assumed in each of the scenarios described below. Here, we note that other possible scenarios exist (e.g. X.25 *tunnelling*, where connectionless packets could be carried within connection mode packets, etc.). However, these are not discussed as they are not relevant for the rest of the book. In both cases, only the B-channels are shown.

CONS across ISDN Figure 3.7 shows a NL relay protocol stack where the CONS is provided across both the LAN and the ISDN. At the network layer both the LAN and the ISDN are assumed to have the X.25 packet layer protocol (PLP). The lower layers are necessarily different. When a new B-channel is set up, the data link must be established over it. The use of multi-link procedures are discussed in Chapter 4 with reference to channel management.

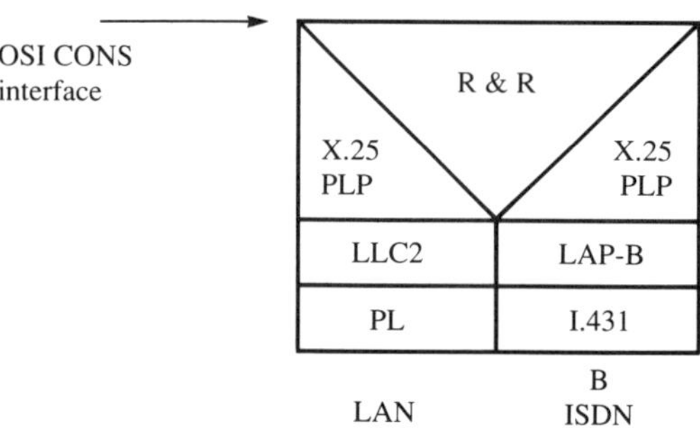

PLP: packet level protocol (IS 8208); LLC2: logical link control type 2 (IS 8802-2)
LAP-B: link access protocol-balanced (IS 7776); PL: physical layer; R & R: routing and relaying

Figure 3.7 A connection oriented LAN–ISDN network layer relay

CLNS across ISDN Figure 3.8 shows a NL relay protocol stack where the CLNS is provided across both the LAN and the ISDN. At the network layer both the LAN and ISDN are assumed to have the CL network protocol (ISO internet protocol). At the data link layer, the ISDN need not have the MAC-level protocol. Indeed, HDLC core protocol could also be used instead of the LLC1. If LLC1 is used, no data link needs to be established after a circuit set-up since this is a CL mode layer 2 protocol. If HDLC is used, however, the data link needs to be established anew with every new circuit set-up. The interaction of CLNP and the channel management are studied further in Chapter 4.

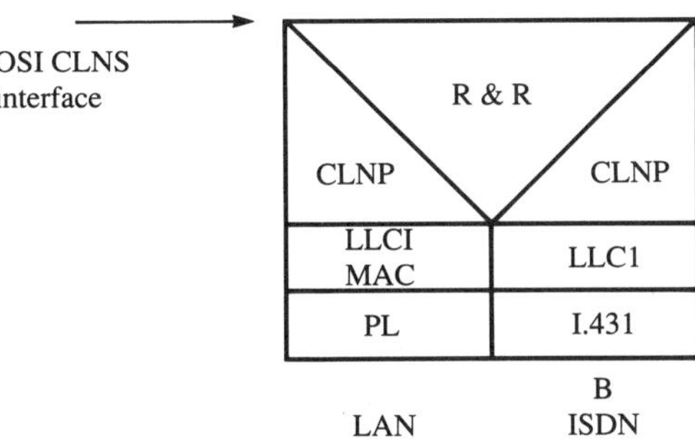

R & R: routing and relaying
CLNP: connectionless network protocol (IS 8473); LLC1: logical link control type 1 (IS 8802-2)
MAC: medium access control protocol; PL: physical layer; I.431: ISDN physical layer

Figure 3.8 A connectionless LAN–ISDN network layer relay

3.4 SUMMARY

Circuit-switched ISDN, providing fast switching, multiple channels, common channel signalling and digital bit pipes, presents a flexible wide area network for LAN–ISDN and LAN–ISDN–LAN interconnection. The user flexibility to choose suitable protocols over raw bit pipes presents a challenge. These protocols do not need to be heavy-weight, owing to ISDN's good bit error rate performance. Network layer interconnection has its advantages in conforming with the OSI-RM and eases the problems of global addressing. Multiple user channels at the UNI precipitate dynamic channel management issues, which are dealt with in Chapter 4.

CHAPTER

FOUR

CHANNEL MANAGEMENT

The OSI Reference Model (ISO 1984) specifies that relaying should be done at the network layer (NL) or below. (NL is favoured for interconnecting on a very large scale.) In the traditional sense, the NL relay has fixed transmission lines (e.g. leased lines as opposed to switched), linking it to one or more destinations (e.g. other relays, LANs or packet-switched network interworking units). However, a NL relay operating in packet mode over the circuit-switched ISDN has multiple channels that can be assigned on demand. This gives an added flexibility to a relay controlling the ISDN access interface so that it can assign bandwidth to requests dynamically. This chapter investigates the implications of this flexibility in terms of channel and bandwidth management and protocols.

Packet operations can be carried out in connectionless (CL) or connection-oriented (CO) modes, as described in Chapter 3. Assuming, for simplicity, CO transport and data link services, two sets of protocol stacks emerge: one with a CO network service (CONS) and one with the CL network service (CLNS). The mechanism of channel management is necessarily different in the two types of network service. In either case, end-to-end associations are established by transport layer entities wishing to communicate with other peer entities. Furthermore, in both cases the establishment of data link service is necessitated once new physical channels are switched in. This is in contrast to the permanent link case where the data link connection is established once at the start-up time and is assumed to exist permanently.

From the channel and bandwidth management aspects, two questions arise:

- When should connections be set up and removed?
- How can the bandwidth of existing connections be varied?

The first question is not as trivial as it sounds. In the first instance, a certain amount of bandwidth must be allocated from the bandwidth pool to a new request. This may be subject to some form of access control. Furthermore, a network layer relay has no knowledge of the end-to-end associations (the duty of the transport layer protocol), and hence it needs to have a strategy for setting up and removing circuit-switched connections. The second question raises the issue of the type of policy to be used to determine the instantaneous bandwidth of a connection. In both cases, issues such as the use of signalling protocols for channel management, practical mechanisms involved in bandwidth variation (e.g. channel aggregation) and the provision of quality of service also arise.

The burst nature of packet data lends itself to statistical multiplexing. This is an inadvertent feature of the CL mode of communication, and the same can be achieved by the use of multiple virtual circuits (VCs) over the same physical link in the CO mode of communication. However, in both cases the bandwidth of the underlying physical link is usually assumed to be fixed. The idea of statistical multiplexing can naturally be extended to the LAN–ISDN interconnection. An added complication here is that the bandwidth of the underlying physical link between two entities may vary (within certain limits). Note that this forms a model similar to the one considered by Harita and Leslie (1979), where variable bandwidth circuit-switched ISDN links using ATM-type packet structures are utilized in interconnecting LANs over a wide area.

In a multi-channel NL relay, there may exist multiple channels (physical or logical) to the same or different destination(s) concurrently. The NL relay dispatches arriving packets (customers) to their destinations according to some specified routing strategy. As mentioned in Chapters 1 and 2, multiple parallel channels or a single superchannel to a given destination may be created. When multiple parallel channels to a given destination exist, the question of packet scheduling to these channel queues arises. As the traffic load (i.e. the aggregate bandwidth requirement) of a set of multiplexed traffic streams may vary over a given period of time, so will the channel (i.e. connection or link) capacity needed to that destination vary. This can depend upon the changes in the number of users[1] or in the types of activities they partake in during the course of an end-to-end communication. It would be wasteful if a large amount of capacity were allocated to the communication needs of a group of users to a given destination when, as a result of changes in their traffic activity, they no longer needed the same high capacity. It would also be unfair to other users who might want to use the same resources. It pays, then, to adjust the allocated capacity (bandwidth) of a channel according to the traffic activity, some specified QoS parameter and predefined cost structures. We call this collection of management activities channel management, as it involves the interaction of several related but distinct disciplines.

The channel management problem is described in Sections 4.1 and 4.2. Sections 4.3 and 4.4 present the bandwidth allocation and management issues. Queuing systems and queuing strategies relevant in channel management are the topic of Section 4.5. Section 4.6 presents a short summary of the possible scheduling policies that could be used in multi-channel interfaces. The effects of CO and CL protocols used over the ISDN B-channels on channel management in a network layer relay are examined in Section 4.7. Section 4.8 presents a short discussion on the management information needed for effective channel management in network layer relays. Cost factors are briefly studied in Section 4.9. Finally, a summary of the chapter is given in Section 4.10.

[1] A *user* can be a real person, an application (process), a transport entity or a packet traffic source for the purposes of this book. Note that user (application) processes are not visible to a NL relay. However, in this context this term refers to transport entities ‘representing’ these users.

4.1 THE CHANNEL MANAGEMENT PROBLEM

Frequently in computer network analysis, issues related to each other are, out of necessity, studied and analysed separately for ease of approach. Examples are bandwidth allocation, bandwidth management, scheduling,[2] design and control of queues, channel assignment and transmission control. However, these issues seldom occur in isolation, and they need a unified approach to achieve the best practical understanding. Channel management is such a problem, as it encompasses several distinct but interrelated sub-problems. Furthermore, the decisions at policy level usually affect the performance at different levels of the communications hierarchy by varying amounts, and their collective effect also needs to be investigated. Some definitions for the above issues are as follows:

- *Bandwidth allocation (BA)* This relates to the problem of partitioning the bandwidth resource at a communications facility among a group of homogeneous or heterogeneous (in terms of their arrival rates, service/holding time distributions, bit rate, queuing/blocking requirements) users. Usually, it is assumed that each user has a static requirement, specified at the start of its communications activity.
- *Bandwidth management (BM)* This concerns the problem of managing the bandwidth of an individual user or a group of users as its/their requirement(s) change over time during a connection. Hence it is assumed that a bandwidth allocation strategy already exists and that some amount of bandwidth has been assigned to this/these application(s). The bandwidth management policy then operates within the confines of the BA strategy such that the bandwidth of a given channel is varied (increased/decreased) accordingly.
- *Scheduling* This describes the type of policy that is used for dispatching packets addressed to a given destination when there exist multiple parallel channels with individual queues. This problem does not exist if only one packet (user/customer) queue exists per destination.
- *Design and control of queuing systems* Classical queuing theory deals with queuing systems in a *descriptive* way. Dynamic control of queuing systems necessitates a *prescriptive* approach. Most real-life situations are not static, and the control of queuing systems as a discipline is of prime importance in channel and bandwidth management.
- *Channel assignment and transmission control* Channel assignment describes the procedures for assigning channels to incoming requests. It includes channel hunting and channel de-assignment for cost minimization. Transmission control involves the use of the signalling procedures for link establishment and removal and data packet handling (transmission and reception).

In this chapter, we concentrate on the dynamic channel management problem in network layer relays in general and in the connectionless NL relay in particular. Some of the issues also apply to equipment directly interfacing the ISDN (rather than through a relay). The control of queuing systems comes into play in the design of bandwidth control strategies. Some aspects of the bandwidth allocation and dynamic bandwidth management issues are also considered in relation to the channel management problem. Scheduling issues are discussed in relation to the management of multiple parallel channels between two communicating end-points. Channel assignment and transmission control are also relevant to the discussion; however, these issues are discussed in Chapters 5 and 6.

[2] Here, *scheduling* refers to *packet* scheduling. This term is preferred to 'routing' as the latter has a special meaning in the N-layer.

4.2 DEFINITION OF CHANNEL MANAGEMENT

In designing NL relays between LANs and the ISDN, it is necessary to manage the ISDN resource in the most effective way in order to meet several conflicting criteria. Bandwidth utilization, user-perceived response, delay, throughput and real communication costs must all be considered. Therefore, the main objective of channel management may be defined as the optimization of bandwidth utilization (hence the reduction of costs) for a given set of users with a given quality-of-service (QoS) distribution. This applies to multiple distinct channels (or a single channel with controllable multiple basic bandwidth units) and describes their management.

A collection of policies regarding bandwidth allocation, bandwidth management, scheduling, channel assignment and transmission control forms a channel management (CM) strategy. A CM policy can be implemented by having controls at two distinct levels. First, the bandwidth allocation policy level determines how the available bandwidth is distributed among the different requests and hence provides an admission (access) control policy. Second, the bandwidth management policy level determines how the bandwidth of an existing channel should be varied as the traffic characteristics vary in time. In the case where multiple parallel channels with individual queues exist to a given destination, a third level of control is available. This is the scheduling policy level, which determines how the traffic is 'routed' to different parallel channels established to the same destination. The scheduling policy in turn provides a handle on the queuing policy to be adopted for the outgoing links. The queuing policy may be QoS-associated.

Thus, channel management encompasses the bandwidth allocation (BA), bandwidth management (BM) and scheduling policies. The BA policy is needed to prescribe from the outset the 'rules' of the game, that is, how much bandwidth each user class or group may have, while the BM policy determines how the bandwidth is varied (increased or decreased) from the initially allocated amount within the limits prescribed by the BA policy. The scheduling policy determines how the packets are 'routed' to different channels connected to the same destination (and forming part of the same link) in order to achieve some QoS or performance criteria.

In the rest of this book, we will concentrate on the channel management problem, in particular the channel set-up/disconnect and the dynamic bandwidth management issues. But first, discussions on bandwidth allocation, control of queuing systems and scheduling are presented in order to highlight their relevance.

4.3 BANDWIDTH ALLOCATION

Bandwidth allocation policies determine how the total available bandwidth of a communication channel is allocated (partitioned) to multiple user (customer) requests. It is also sometimes referred to as circuit-channel access control policy since the total bandwidth is partitioned into 'channels' (Kraimeche and Schwartz 1984; Schwartz and Kraimeche 1982). Most BA schemes are based on a *prescriptive* method. This means that each user (user process, or *protocol data unit*) requests a certain amount of bandwidth at the start of each session, and is content with that amount of bandwidth for the duration of a session. Also, there are limits to the amount of bandwidth a class of users is allowed to occupy in a situation of bandwidth contention with other classes of users.

An important design aspect from the BA policy point of view is the composition of the users. Generally, they are classified into two types: blockable and queueable (class 1 and class 2, respectively). These two classes are sometimes referred to as circuit and packet types of user. However, strictly speaking this is not true, since packetized voice would still be in class 1 despite its packetization. This terminology merely indicates the type of service required by a class of users. Hence the circuit users are assumed to be operating on a blocked-calls-lost principle, while the packet users are assumed to operate on a queuing principle. Different BA policies can be devised on the assumptions of homogeneous or heterogeneous blockable/queueable users. Traditionally, the circuit and packet traffic have been treated separately. Each type of traffic would use different types of networks and hence switching. In the ISDN era this is no longer true. Switching systems handling both the circuit and packet types of users are sometimes called *hybrid switching* systems.

One way of dealing with mixed traffic types is to separate the available bandwidth into the 'circuit' and 'packet' groups. In a TDM frame, this can be done by partitioning the available time slots, N, into two groups, N_1 and $N_2=N-N_1$ slots. A boundary is thus formed between the class 1 and class 2 traffic types where N_1 slots are assigned to the blockable class while N_2 slots are reserved for the queueable class of users. This boundary can be fixed or movable. In the movable boundary scheme, class 2 packets may occupy any of the unused N_1 class 1 slots, with a proviso that any arriving class 1 call can pre-empt a class 2 packet occupying these slots. The movable-boundary hybrid switching scheme is found to be superior to the fixed-boundary scheme in its improved time delay performance for the queued class of users. However, care must be taken with the movable strategy. Any attempt to increase the class 2 user throughput above that which is normally available by N_2 slots reserved per frame, by the utilization of class 1 user slots released on the average, may lead to exceptionally long packet queues (Schwartz 1987; Schwartz and Kraimeche 1982).

In either case, parts of the total bandwidth may be reserved as a general shared pool, which could be equally accessible by all user classes or could have some form of restricted access superimposed. The prescriptive nature of these policies guarantees a certain QoS, or resource utilization, and fairness to contending users (Schwartz and Kraimeche 1982).

In the case where only blockable (circuit-switched) type traffic is available, bandwidth allocation can be done on the homogeneity/heterogeneity basis. Some of the previously studied strategies for BA in circuit-switched systems include: complete sharing (CS), complete partitioning (CP), restricted access (RA) and restricted access with priority (RA-P) (Kraimeche and Schwartz 1984). In complete sharing, no explicit control is exerted on the access of the K circuit streams and an arriving message is rejected only if it requires a bandwidth greater than the available amount. In complete partitioning, the total bandwidth is divided into K separate portions such that each circuit stream has its own dedicated bandwidth. In the restricted access type of policies, the users are grouped into K user types where each type is characterized by its bandwidth requirement; RA policy then restricts the resource occupancy of users from each group, hence reserving part of the total bandwidth to each group. An RA policy with priority permits users of lower bit rate to access the bandwidth reserved for users of higher bit rate which have a right to pre-empt lower bit rate users.

It is known that, for homogeneous circuit traffic (i.e. same bit rate and holding time), the CS policy yields best throughput (Kraimeche and Schwartz 1984). For non-homogeneous traffic with a large spread in its bandwidth requirements (i.e. bit rates), a hybrid CP policy, where the bandwidth dedicated to a given group is completely shared by user types in that group, outperforms any other simple policy. It is also known that a RA policy augmented by a priority scheme provides improved performance over other policies described (Kraimeche and Schwartz 1984).

4.3.1 Packet data over circuit-switched connections

Bandwidth allocation in the above sense can be achieved only if the bit rate requirements of users/applications are known in advance. That is, when such users arrive at a resource facility, they request a certain amount of bandwidth. They could then be allocated their requested bandwidth according to the BA policy implemented by the resource manager.

This form of BA is suitable only in the CO mode of interworking as the QoS parameters (indicating bandwidth or data rate required as well as the acceptable delay and throughput parameters) for each CO session are agreed upon between the user[3] and the resource managers at the intermediate nodes of a communications link at the start of an association during the set-up procedure.

In the case of CL working, the QoS is not negotiated at the start of a communication but is determined in an *a priori* fashion between the users and the service providers (ISO 1987b). Also, as there is no connection set-up phase in this mode of operation, the bandwidth requirement of an application cannot be determined. Furthermore, the QoS parameters of the connectionless network service do not include bandwidth (see Section 4.7). In CL network layer relays, traffic multiplexing is an inherent feature, although this may be specifically prevented in order to guarantee a QoS. However, the CL networks operate on the best-effort principle and do not guarantee a QoS. When several traffic sources are multiplexed onto a physical channel and the occurrence and duration of these sources vary stochastically, the bandwidth requirement of the channel also varies in a stochastic manner. Under these conditions, the BA policy may actually limit only the upper bound of bandwidth allowable for each channel. It is then left to the interaction of the user traffic and the bandwidth management policy to determine the amount of instantaneous bandwidth allocated to such a channel. In the rest of this study, this form of BA control will be assumed.

4.4 BANDWIDTH MANAGEMENT

This is the resource management function associated with the adaptation of user bandwidth to the needs of user traffic which may vary during the lifetime of a connection. We have already considered scenarios where such variations in bandwidth requirements may occur.

The bandwidth of a TDM access structure is divided into channels, each of which may be mapped to one or more time slots. Each TS can be considered to be a basic bandwidth unit (BBU). As the traffic load of a stream (or a multiplexed stream) increases over time, the bandwidth allocated to this user should be increased in order to provide a constant QoS performance. This bandwidth increase can be put into effect by associating more TSs to this channel.

[3] Assuming a one-to-one correspondence between 'users' and 'network connections'.

At the core of the bandwidth management problem is the question of *how* and using *which* type of control this increase and decrease in the channel capacity (bandwidth) should be achieved. Service capacity (rate) control of queuing systems using *threshold* policies based on the system state is well known (e.g. Gebhard 1967; Moder and Phillips 1962; Romani 1957; Yadin and Naor 1963). These policies could be based on single or multiple thresholds and point or hysteresis forms. Figure 4.1 shows a schematic of point and hysteresis threshold forms when the system state used is the instantaneous queue length. A *point threshold* policy implies that a decision to increase the service capacity of the server is made as the system state value 'crosses' the threshold value from below; a decision to decrease of bandwidth is made as the system state value crosses the threshold from above. A *hysteresis threshold* policy has two such levels. A decision to increase the bandwidth is made when the system state value exceeds the high (upper) threshold. This capacity is held until the state drops back again to the low (lower) threshold value, whence a decision is made to decrease the bandwidth. This is called a *two-phase hysteresis*. More complex schemes with a multiple threshold pairs (*multi-phase hysteresis*) that may or may not overlap can be devised (Yadin and Naor 1967). Also, the level of control could be based on either bi-level or multi-level types. In the bi-level service rate control, the system can have one of two service rates, μ_1 or μ_2. Whichever is selected depends on the state of the control variable. In the multi-level service rate control, the system can have different service rates selected from a set of μ ($\mu_1, \ldots, \mu_N$).

Point threshold control is a special case of the hysteresis control in that the two threshold levels are coincident. This form of control is better at conserving resources but can produce a very high number of switching transients when the system is operating around the threshold value. Another special case of the hysteresis control is one where the lower threshold is set at zero level of the control variable (Gebhard 1967). These are discussed further in Section 4.5.2, with a full taxonomy for threshold control given in Chapter 8.

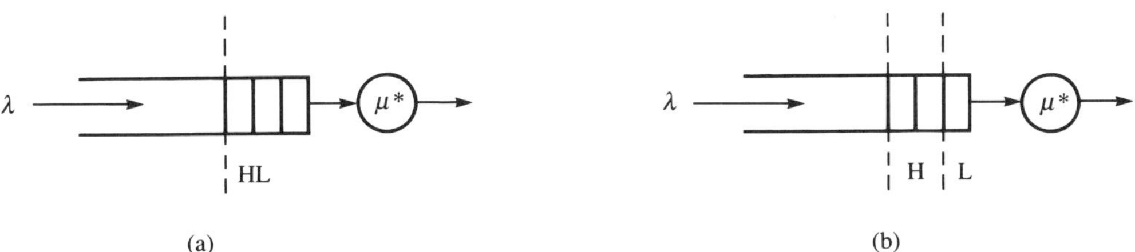

H: high threshold; L: low threshold; λ and μ : mean packet arrival and service rates, respectively
HL: point (high–low) threshold; μ *: variable capacity server

Figure 4.1 An example of (a) point and (b) hysteresis thresholds

4.4.1 Control variables and metrics for BW management

Various control variables (based on system state) and metrics are available for service rate control in queuing systems. These include the instantaneous or averaged values of the number of customers in the queue (*queue length*) or the system, the amount of unfinished work in the system (i.e. the actual number of bits to be transmitted rather than the number of packets in the system) and the packet arrival rate (or the traffic bit rate).

Figure 4.2 shows the control variables and modules for dynamic bandwidth management in a multi-channel packet communications facility as applied to an aggregate packet stream.

The bandwidth management unit (BMU) monitors the system state periodically or on each new packet arrival. This includes the queue length and the data input rate on aggregate traffic streams destined to the same remote location or device (e.g. a network layer relay). The BMU signals the channel allocation unit (CAU) for the implementation of any bandwidth increase/decrease operation. A packet or job scheduler operates to determine which channel (indicated as 1, 2, . . . , N BBUs) receives the next packet or byte. The scheduler operates according to an assigned scheduling policy.

For multiple aggregated traffic streams, the queue and scheduler components are replicated. The output transmission rates of individual aggregated channel groupings must also be input separately into the bandwidth management unit.

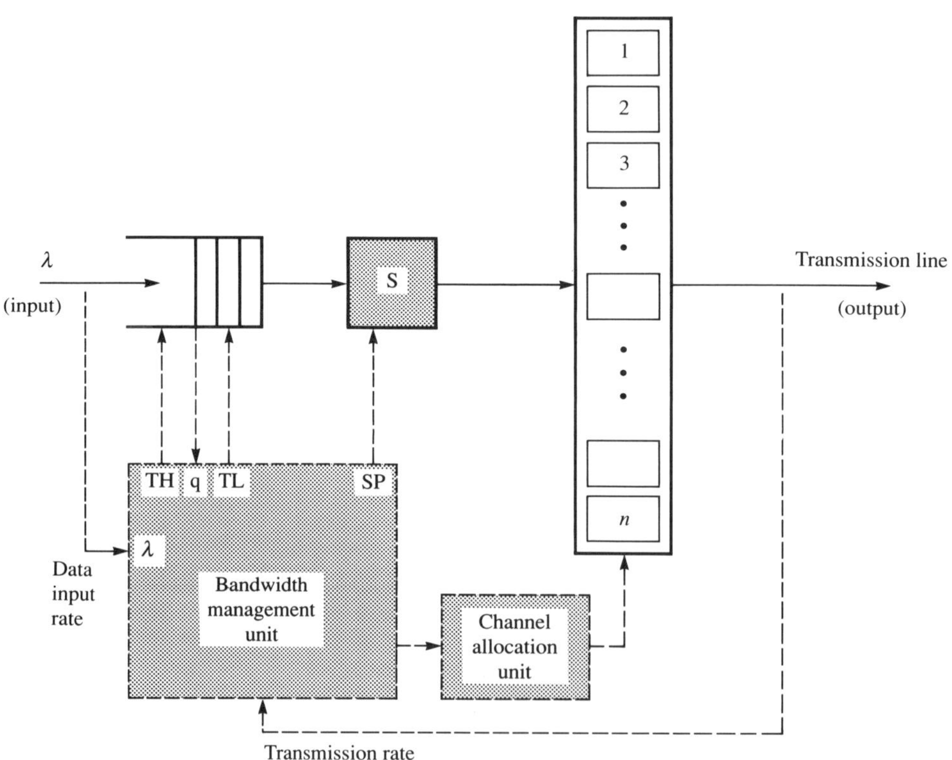

TH: threshold high; TL: threshold low; q: current queue size
SP: scheduling policy; S: packet scheduler; λ : packet arrival rate

Figure 4.2 Components of bandwidth management in a multi-channel packet communications facility

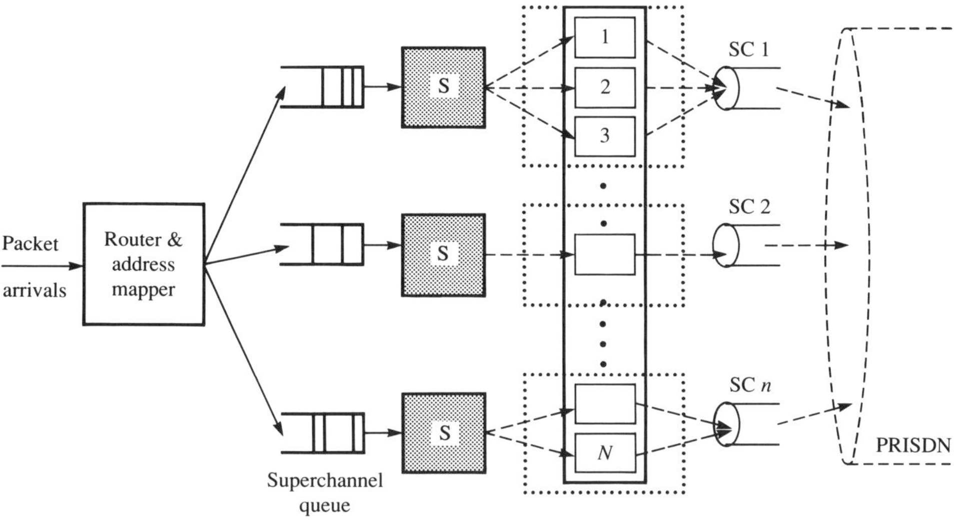

Figure 4.3 Formation and use of multiple superchannels in a network layer relay

Figure 4.3 shows the formation of multiple superchannels in a packet switch or network layer relay. The BMU of Fig. 4.2 is applied to each aggregated packet stream in superchannel queues.

In queuing theory, bandwidth management corresponds to queuing systems with removable or variable capacity servers. This class of queues is described next.

4.5 QUEUING SYSTEMS FOR CHANNEL MANAGEMENT

The basic and primary-rate ISDN interface channel structures were discussed in Chapter 2. Three distinct queuing models (see Fig. 4.4) can be developed depending on how the n time slots (each capable of carrying a byte of information per frame) in a TDM frame are utilized. Assuming that the interface is composed of $n \times$ BBUs (i.e. 64 kbps) and no QoS queuing[4] is undertaken, these are:

- *Model 1* n parallel identical servers each with its own queue. This gives an $n \times$ A/B/1 queuing system where each server gets its own packet to serve.
- *Model 2* n parallel identical servers with a single queue. This results in an A/B/n type queuing system where each server gets its own packet to serve.
- *Model 3* Variable capacity server with a single queue. This results in an A/B/1 queuing system with capacity $n \times$ BBU, where each TS serves a byte of a packet in turn.

[4] In QoS service queuing, multiple queues may be associated with one or more servers (Prue and Postel 1988).

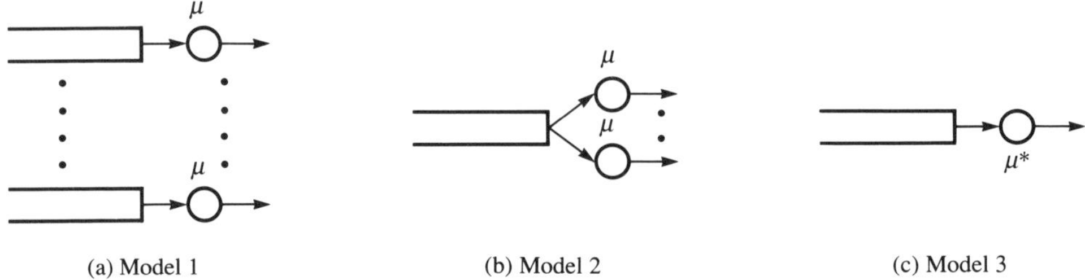

μ:* variable capacity server

Figure 4.4 Alternative queuing models

Variable bandwidth models Variable bandwidth queuing models can be obtained from the above basic models by having n as a variable. A *logical* channel is assumed to be formed by the association of physical channels. In model 1, the logical channel bandwidth can be varied by the addition or deletion of servers and their queues. When multiple servers with individual queues to the same destination exist, the problem of packet scheduling arises. In model 2, a BBU (time slot) can be associated with, or dis-associated from an existing logical channel queue, thereby providing a variable bandwidth model. In model 3, each added TS is closely integrated with the existing TSs, thereby effectively increasing the server capacity. Server capacity can be reduced simply by dropping some TSs from the logical channel.

In practice, models 1 and 2 present a re-sequencing problem at the destination since packets can be reordered by this form of transmission. Response time performance of model 3 is superior to model 2, which itself is superior to model 1 (Kleinrock 1974, 1976). However, model 3 presents a time slot sequence integrity problem. These issues, and the practical ways of implementing variable bandwidth channels in an ISDN interface, are described further in Chapters 5 and 6. In Chapter 5 a dynamic channel management architecture is presented for models 1 and 3. This architecture can easily be modified to address model 2.

Variable bandwidth control for logical channels For models 2 and 3, this can be achieved simply by the use of *threshold* policies described in Section 4.4. In model 1 two options exist regarding the use of threshold policies for service capacity control: the use of a logical queue length (sum of individual queue lengths) as the control variable, and the use of separate thresholds on each queue. The latter could be combined with the scheduling policies but inevitably leads to a more complex control strategy (see also Section 4.6).

4.5.1 Queuing systems with removable servers

Queuing systems for channel management in ISDN form a class of models that includes addable/removable servers. In general, the performance of these systems can be studied by considering queuing systems in which the service capacity varies as the unfinished work varies. The decision mechanism for varying the channel capacity could depend on the buffer occupancy, the cost incurred under various conditions and cost structures in the model. The class of queuing systems with removable servers has been studied by various researchers (see also Section 1.6). A comprehensive study on optimal design and the control of queues is by Crabill *et al.* (1977). One of the earliest studies on variable bandwidth queuing systems is by

Moder and Phillips (1962). Later work includes the paper by Yadin and Naor (1963). This study considered a single server that could be either present or absent. Formulae were derived for the length of busy periods and the proportion of time that the server was in operation, closing down, starting up or idle. In a later paper (Yadin and Naor 1967), a system in which the service capacity could vary was considered. As the number of customers waiting increased, at various thresholds the server capacity would increase; the service capacity would be decreased as the number of customers decreased past different thresholds. In the latter paper, the addition of new service capacity was made instantaneously. Several other authors have dealt with variable-rate servers, but they have all assumed zero dial-up (or set-up) time (Bell 1980; Levy and Yechiali 1975), which is not realistic in telecommunications.

In a study by King and Shacham (1986), two models of buffers with dial-up capability are analysed. In the first model, a buffer that has a permanent and a dial-up channel connected to its output is analysed. The dial-up channel is added into the system forming an M/M/2 queuing system whenever the occupancy (the number of messages in the queue) increases above some threshold. When the buffer is empty, the overflow channel is released. A second model considers a system that consists of two buffers that use dial-up channels to communicate. Each node that has its buffer occupancy above the threshold dials up a channel. A connection, once established, is full duplex, allowing data to flow in both directions. Thus, when the first dial-up is complete, the other buffer, if it is in the middle of a dial-up, aborts that process and uses the existing channel to send messages. The channel is released when both buffers are empty.

In this class of queuing systems it is known that a *threshold* policy, i.e. initiating connection establishment on the arrival of the Kth message to the buffer, is optimal (Heyman 1968). The problem then becomes one of finding the optimum threshold that will minimize the total cost owing to channel set-up and to delays incurred in the queue. This threshold depends on system configuration, rates of service and arrival and various costs.

4.5.2 Threshold control of server capacity

It is known that the M/M/1 queue has a monotone hysteresis optimal service rate (or capacity) control policy when switching costs exist and a monotone optimal control policy without switching costs (Lu and Serfozo 1984). The hysteresis control form, where the switching costs exist, arises from the need to minimize the number of switchings and hence make it economical once an increase in service rate is achieved. The hysteresis service rate control, where the switching occurs instantaneously, has been studied by Gebhard (1967) and Li (1988). The case where the switching takes a time that is exponentially distributed has been studied by Harita and Leslie (1989). All of these studies consider only a bi-level change in the service capacity (e.g. $\rho_0 = 2.0$ and $\rho_1 = 0.5$ where ρ = utilization factor = λ/μ). In Chapter 8, several heuristic *multi-level* hysteresis policies are proposed for an M/M/1 queuing system and their performance is analyzed using simulation models.

4.5.3 Queuing systems with fluctuating parameters

Many real-life queuing systems, such as communication channels, exhibit random fluctuations in their message arrival rate. The statistical variations of the offered traffic load to such systems introduce system design requirements that cannot be accounted for through the use of a conventional time-homogeneous traffic model (Zukerman and Rubin 1986). The ISDN relay described in this book is no exception.

Non-homogeneous queuing systems so far analyzed in the literature can be divided into two main categories. In the first, the arrival process is assumed to be non-homogeneous Poisson with rate function $\lambda(t)$, where $\lambda(t)$ is a deterministic function of time; and in the second, the arrival process is assumed to be a doubly stochastic Poisson process; i.e., the arrival rate function, $\lambda(t)$, is itself a stochastic process (Zukerman and Rubin 1986). The latter approach is relatively new, and the first work was published by Yechiali and Naor (1971) looking at a special case where the arrival and service rates to a single M/M/1 queue are step function stochastic processes. Under this assumption, the queuing system can be in one of two modes, 0 or 1. When the system is in mode i, the interarrival and service times are exponentially distributed with parameters λ_i and μ_i, $(i = 0,1)$, respectively. The time interval during which the system functions at level i $(i = 0,1)$ is exponentially distributed with parameter γ_i. Further extensions to the theory have been achieved by Neuts (1971) by assuming a general number of levels and general service time. Generalization of the above case to more than one server is developed by Zukerman and Rubin (1986).

In the case where the arrival process of a queuing system can be in one of two modes (i.e. represented by a two-state Markov process) these modes could represent *light* and *heavy* loads. This leads to the definition of two further study models: model I, where a single-mode arrival process is assumed, and model II, where a two-mode arrival process is assumed. Model I is dealt with in Chapters 7 through 10 in detail, while Model II is considered beyond the scope of this book.

4.5.4 Queuing strategies

The factors affecting the queuing strategies in a relay within a heterogeneous environment are somewhat different from those of the data-only environment that we are considering. In the heterogeneous environment the relative mixes of different traffic types as well as their service requirements in terms of switching are more important. The delay on synchronous services like packet voice and video must be bounded. Error rates become more important if some form of compression has been used. In the data-only case, delay is tolerated to a certain extent but reliable delivery is necessary. The factors described below are considered to be most relevant to this book.

Application types Here the difference is between the interactive versus non-interactive applications. The problem is that the network layer has no knowledge regarding the type of higher-layer protocol data units carried within a packet. It is up to the transport service to determine the QoS parameters that will be included in the network service data units. Also, different priorities may be assigned to different application types.

Quality of service The determination of the quality of service for CO and CL modes of transmission are different from one another. Here we consider only the CL case. The OSI connectionless network service (CNLS) definition describes the associated QoS parameters (ISO 1987b). The DARPA IP type of service parameters (Postel 1981a) are similar.[5] The transit delay and priority QoS parameters are most important in the context of channel allocation strategies.

[5] QoS parameters included in CLNS are: transit delay, priority, protection from unauthorized access, source routing, residual error probability, and cost determinants. QoS parameters included in the DARPA IP are: precedence, delay, throughput and reliability. However, these have qualitative attributes, e.g. high or low throughput.

The transit delay QoS parameter will indicate how that packet should be treated in terms of queuing for output within the CL network relay. For low transit delay requirements, more channels may need to be opened in order to satisfy the network service user, although the CL mode of operation only endeavours best effort in delivering a particular packet to its destination. Note that the OSI CLNS does not specify a QoS parameter for throughput.

A queuing algorithm for type of service in TCP/IP networks is developed by Prue and Postel (1988), where multiple channel queues are assigned to a server. The server spends a proportion of time serving each queue. The time spent on each queue is linked to the number of queues, the service rate and the QoS requirements. However, for simplicity we do not implement the full QoS queuing in our channel management plan. The only QoS parameter considered is the 'delay' parameter. Hence we assume that all the packets to the same destination are multiplexed into a single queue, irrespective of their QoS requirements.

Priorities and pre-emption The use of priorities for different classes of applications has not been widely implemented in CL networks although some work has begun (Prue and Postel 1988). However, the QoS parameter field is available for this purpose. It could possibly be used as a basis for giving higher priorities to packets from certain types of application. For example, an interactive application such as a file access can be given higher priority than that for an electronic mail application. Pre-emption suggests that a higher-priority packet could pre-empt a lower-priority packet in order to get speedier service.

The issues of priorities and pre-emption are beyond the scope of this book and hence will not be considered any further.

Costs Opening a new circuit for a burst arrival of packets may be expensive if that channel is not going to be utilized after the burst has been cleared through the system. Therefore, ways of determining the best threshold for new circuit set-up on possibly multiple queues (e.g. model 1) to the same destination need to be found.

4.6 SCHEDULING POLICIES

By 'scheduling policies' we mean the way in which the individual packets are 'assigned' to different physical channel queues of a *logical* channel, and as such this applies to model 1. In queuing theory terminology, this entails the scheduling of a new job (customer, packet) to a server (channel or its queue). Some well-known scheduling policies are: 'round robin', 'random' and 'shortest queue' policies (Bhattacharya *et al.* 1987; Ephremides *et al.* 1980; Nelson and Philips 1989; Tran-Gia and Rathgeb 1987). In real data communication interfaces the size of the channel queues are bounded; therefore, scheduling policies like 'sequential' and 'alternate sequential' policies can also be used.

Under the round robin policy, each queue receives a packet in turn within a strict order. Random policy queues packets in a random fashion to the channels. Under the shortest queue policy, an arriving packet is placed in the queue with the shortest length. A variant of this policy, the 'shortest remaining work policy', places the arriving packet in the queue with the least unfinished work (i.e. fewest total number of bits waiting to be transmitted, rather than smallest number of packets in the queue).

The sequential policy applied to a two-channel system would first start with channel 1 and schedule all incoming packets to its queue until it reaches the maximum queue size or a set threshold value. The second channel would be used as an overflow channel only, so that, whenever the channel 1 queue size fell below the set value, it would get the following packets as described above. The advantage of this method is that it will use the overflow channel(s) only when it can handle no more packets. An *idle-timeout mechanism* can be used to remove (switch off) the unused overflow channel(s). The alternate sequential policy behaves like the sequential policy except that it alternately fills the channel queues to their limit.

4.7 PROTOCOLS OVER THE ISDN B-CHANNEL

As was mentioned in Chapter 3, the two modes of interworking at the network layer—CO and CL—lend themselves to different protocol stacks over the ISDN B-channels (see Section 3.3.5). Here, we investigate the effects of the mode of operation. The choice of protocol stacks to be used over the ISDN B-channels and their behaviour are important in the type of channel management schemes to be adopted (Knight *et al.* 1987). This is discussed below with reference to CO and CL network services (CONS and CLNS). It is assumed for simplicity that a CO transport service (COTS) is used in both cases.

The channel and bandwidth management aspects mentioned at the beginning of this chapter are relevant here. This section analyses only the protocol interaction with the connection set-up and removal operations. The question of bandwidth variation using threshold control applies to the CL relay more readily than to a CO relay, since in the former multiplexing of packets from several TP entities into one channel queue is assumed. This also relates to *window*-based flow control issues in transport protocols (TPs), since the bandwidth used by a TP would be controlled by this mechanism when operating over a connectionless network service (e.g. TCP/IP). In the CO relay case, the various virtual circuits (VCs) multiplexed over a single physical circuit produce a queuing model where a server is serving multiple packet queues, one per VC. A detailed study of that model is beyond the scope of this book.

4.7.1 CO relay

A CO service data unit for call set-up addressed to a remote host will be recognized and the next hop along the route will be determined. Assuming that there exist no circuits to the destination, the relay will set up in a circuit-switched path to the next hop, using its ISDN layer 3 call set-up procedures on the D-channel. This will then be followed by the normal X.25 call set-up procedures on the B-channel. Following the set-up of a virtual circuit, the data transfer will take place. In this case X.25 calls will be served to completion, i.e. until one of the parties terminates the connection.

If other users wish to communicate to the same destination relay while this circuit is set up, the same physical circuit could be used for the new transfer by setting up a new VC which will be multiplexed over the existing physical circuit. However, if the QoS parameters exchanged at the beginning of the session demand a full 64 kbps service, then a second B-channel will be used for a new call set-up (provided one is free); otherwise, a *disconnect* packet will be returned.

An alternative mode of working is to use multiple *physical* ISDN circuits to provide a larger bandwidth for applications requiring it. In this case, the X.25 multi-link procedure (MLP) could be used over single-link procedures (SLPs) running on each physical circuit (Deniz and Knight 1989b).

The above method of usage means that the slow terminal or file transfer traffic could be multiplexed over one ISDN circuit-switched connection, whereas the rapid response requirements of a file access would be met fully by avoiding further call multiplexing on that physical circuit. In the case of fast file transfers, MLP could be used over multiple circuit-switched connections to provide even larger bandwidths.

In order to achieve an effective channel management, therefore, the relay needs to know about the parameters of each connection as well as other state-dependent information—for example the delay and throughput requirements of each VC connection, the address of the destination relay and the B-channel(s) to which it is connected. Also, a binding of VCs to 'physical' ISDN circuits is necessary (Deniz and Knight 1989a).

4.7.2 CL relay

In the case of connectionless relay, the problems with the circuit set-up and removal include the following:

1. Applications running on LAN hosts may need to establish an association across the ISDN. Ideally, an ISDN circuit should be established as soon as the association requires it, and should be terminated as soon as there are no associations extant. If the relay is a network layer relay, it has no idea of the progress of the association, its duration, or the number of packets to be transferred; it needs to have a *time-out* mechanism which will close the connection down if no more packets arrive during that duration. There is obviously a danger of closing the circuit while the association is merely quiescent.
2. As the end-to-end connection are initiated and maintained by the transport protocol, its features become important in the above case. For the practical case of ARPANET transmission control protocol (TCP) over internet protocol (IP), it is necessary to be aware of the time-out values of the implementations. The main time-outs in TCP (Postel 1981b) are the user time-out (UTO), retransmission time-out (RTO), time-wait time-out (TWTO) and the maximum segment lifetime (MSL) which indicates the lifetime of any segments remaining in the network. These are discussed below.

In most implementations the retransmission time-out is determined by the hosts dynamically. This means that the relay has no idea of its value for that connection. Furthermore, a transport entity will usually continue to retransmit a transport protocol data unit (TPDU) that requires an acknowledgement for a number of times N; hence ($N*RTO$) is the persistence value. A transport protocol (TP) connection is closed if the inactivity time-out (e.g. TCP's UTO) value is exceeded. It usually indicates failure of the supporting network connection or the remote transport entity. In order to prevent expiration of the remote transport entity's inactivity timer when no data is being sent, the local transport entity sends acknowledgement TPDUs (ACKs) at suitable intervals. This (usually implementation-specific) inactivity time-out value may be used in calculating an ISDN channel's no_activity time-out (NTO) within the relay. Hence the circuit-switched channel(s) could be closed soon after such an event in the absence of traffic on other TP connections. It is also plausible that, in the absence of data TPDUs, if the 'inactivity' ACKs sent by each TE are generated at reasonably long intervals, one may close the channel while there is no real data traffic, and occasionally open the channel to transmit these ACKs and maintain the two end TP connections. A similar argument can be proposed for the sufficiently long RTO values.

To summarize, it is necessary to close unused connections, while care must be taken not to be too quick for fear of closing down an ongoing transmission. Selection of the channel NTO value is of great importance in this respect. It can be seen, therefore, that there is a need for studying the effects of the channel management schemes on the higher-layer protocol operation; herein lie the trade-offs and the route to an efficient relay implementation.

4.8 CHANNEL MANAGEMENT INFORMATION

This is an important issue, as the detail of state information available to the channel management function will determine the efficiency and performance of any channel management policy. Another important issue is the timing of the availability of such information. If there is a delay in state information updating, the actions taken can be detrimental to performance.

From the foregoing discussions, it can be concluded that the following information is necessary for channel management: status of each physical channel available at the interface (e.g. free or used) and the destination address connected; bandwidth associated with each channel in use; and transmission (and if necessary reception) queue sizes (or a variable keeping track of the number of packets in the system—one per server/queue pair). Additionally, if the traffic arrival rate is to be used as the state variable for control, a rate measure per logical channel needs to be kept and updated. It is also necessary to keep a timer (or time-out) variable per logical channel in order to decide the removal of an unused channel. Furthermore, QoS parameters associated with each logical channel need to be kept if certain performance values are to be provided. This field would be updated by the QoS parameters indicated by the network protocol packets.

These issues are studied further in Chapter 5 with regard to the channel management architecture, where a practical method of extracting and using such information is designed. Channel management in connectionless LAN–ISDN relays is also investigated.

4.9 COST FUNCTIONS

The form of cost functions to be adopted in evaluating a queuing system is important in determining the relative importance of cost factors as well as performance measures. Cost factors can be separated into the telecommunications costs incurred in accessing the bandwidth, and the user-perceived response in terms of delay and throughput. In the latter case, by penalizing the increase in delay or reduction in throughput arising from the implementation of a certain management policy, a basis for comparison between different policies can be achieved. However, combining the two cost factors is not straightforward. These issues are discussed further in Chapter 8. Here, a brief discussion is given on the two cost factors.

4.9.1 ISDN tariffs

The tariffs in circuit-switched ISDN are discussed in Section 2.10 above, which points to a tariff structure based on the same principles as for telephone calls; the cost incurred is calculated by the duration and charging index (dependent on the time of day and the distance). It is assumed further that a *stepwise* cost structure exists, i.e. that the charging is made at the beginning of each charge-period. This means that, once a call is established, it makes more sense to keep a channel connected until the end of the current charge-period, unless other messages are waiting for the channel to be released for access. Therefore, tariff cost minimization calls for cutting a circuit connection as short as possible (at the boundaries of charging periods) in order to minimize tariff costs.

4.9.2 Delay cost factor

A cost factor depending on the waiting time in a transmission queue or the sojourn time of packets can be used to add the delay penalty into the cost function. This cost factor can be given a linear or non-linear weighting in order to see the effect on the policy performances.

4.10 SUMMARY

This chapter has presented an examination of the channel management problem. Associated problems of bandwidth allocation, bandwidth management, control of queuing systems, factors affecting queuing strategies and scheduling have also been discussed. It is shown that the multiple channels at the ISDN user–network interface can be used in a variety of ways which could be modelled by alternative queuing systems. On the protocol side, the CO and CL mode services over the ISDN B-channel were considered with particular reference to the channel management problem. The chapter ended with a description of the types of information needed for effective channel management and the cost factors affecting the performance of management policies.

CHAPTER

FIVE

CHANNEL MANAGEMENT ARCHITECTURES

Dynamic channel management (DCM) is needed in order to manage dynamically the multiple channels that are available at the user–ISDN interface. An increase or decrease in the bandwidth of a given logical channel formed by the aggregation of several physical channels (or time slots) can be achieved according to the requested quality of service—e.g. delay, and traffic patterns for providing the QoS and minimizing operational costs.

This chapter describes a dynamic channel management architecture (DCMA) based on the *status table* approach (Deniz and Knight 1989b), whereby the information relating to the individual call, channel statuses and the ISDN interface is kept and updated within the interface by a channel management function. The model described applies to connectionless network layer (CL-NL) LAN–ISDN relays in general and to the Ethernet–PRISDN CL-NL relay in particular. By applying suitable modifications, the architecture can be ported to connection oriented (CO)-type relays as well as to user equipment operating in either the CL or CO modes and having direct access to a multi-channel user–ISDN interface. Furthermore, the architecture described refers to the packet mode usage of the circuit-switched ISDN. The interface referred to is the primary-rate ISDN (PRISDN) with 30 B-channels, although other multi-channel interfaces are also relevant (e.g. PRISDN with a mixture of B and H channels; BRISDN, etc.).

The DCMA developed in this chapter will be shown to serve two purposes: first, as a general model for implementation within a 'real' LAN–ISDN CL-NL relay to provide dynamic channel management; second, as the basis for a simulation model for measuring the performance of different bandwidth management strategies. The implementation of the DCMA within a simulation model has the advantage of proving the viability of the architecture. The level of detail provided in the DCMA model presented here is necessary for the simulation model since the interaction between different decision processes (including signalling and data) as well as packet scheduling to the channel queues needs to be investigated. The issues of simulation modelling and performance measurements are described in the Appendix I and Chapter 7, respectively.

In Section 5.1, general architectural models applicable to the ISDN user–network interface (UNI) are described. This follows on from the discussion in Chapter 4 on the queuing systems for channel management. Section 5.2 presents the *channel management architecture*, defining its main functional units and describing their operations. Section 5.3 presents an application of the design to the Ethernet–ISDN connectionless network layer relay. Finally, a summary of the chapter is given in Section 5.4.

5.1 GENERAL ARCHITECTURAL MODELS

The main assumptions regarding the architecture design are the existence of:

- Multiple channels at the user–network interface
- Common channel signalling (CCS) facility

In Chapter 2, the notion of *channels* formed by the individual or aggregated usage of time slots (TSs) in a TDM frame was introduced. In Chapter 4, channel management was defined and the pertinent models for different types of usage of these channels were described as queuing systems for channel management. Here, the architectural models for two types of queuing systems will be considered (see Fig. 4.4(a) and (c)):

- The $n \times$ A/B/1 queuing system (model 1)
- The A/B/1 queuing system with capacity $n \times$ C (model 3)

Model 1 represents the case where individual B-channels (time slots) are used in parallel, each with its own queue, and to which packets are 'routed' according to the scheduling discipline used. Each B-channel transmits a complete packet. Packet re-sequencing is left to the transport layer protocol operating end-to-end. In model 3, the additional B-channels that are added to the first one share the load in such a way that each TS transmits a byte of a given packet in turn. This mode of usage necessitates the provision of time slot sequence integrity (TSSI) between the individual TSs on an end-to-end basis (at the physical layer) across the ISDN. This is assumed to exist in the rest of this chapter. The issue of TSSI provision is discussed further in Chapter 6. In this model, the underlying physical channels are transparent to the packets since the aggregated channels form a logical channel. Table 5.1 shows a comparison of the two basic models.

Model 2 results from the A/B/*n* type of queuing system (see Fig. 4.4(b)). In this model, one transmission queue per superchannel is needed, and each server (time slot) must have a buffer space for one NL packet while transmitting it. Each server interrupts or raises a flag at the end of its busy period to indicate its idle state so that the next packet can be assigned to it for transmission. Also, the TSSI is not necessary if the transport protocol is used for re-sequencing of network layer packets.

The architectural requirements of model 2 are very similar to those of models 1 and 3. The similarity with model 1 arises since each TS is also a physical channel (i.e. a B-channel), and a superchannel is formed by the association of different B-channels into a larger transmission group. In contrast, model 3 may or may not be composed of individual physical channels: in the *aggregation* method of superchannel formation each TS will correspond to a

Table 5.1 Comparison of $n \times$ A/B/1 and A/B/1 — $n \times C$ queuing models

Features	$n \times$ A/B/1	A/B/1 — $n \times C$[a]
Scheduling	Packets are scheduled to the individual server queues	Not needed since only one queue per 'superchannel' exists
TSSI	Not necessary since transport protocol is used for packet re-sequencing when CL-NP[b] is used	Needed since byte reordering may result. Packet re-sequencing is not needed if TSSI is provided
Response time	Inferior	Superior[c]
BW management	Complicated	Simpler
End-to-end superchannel identification	'*A priori*' or as needed	'*A priori*' or as needed

[a] C: capacity of one time slot (i.e. one basic bandwidth unit—BBU).
[b] *CL-NP*: connectionless network protocol.
[c] Superior response time in the A/B/1—$n \times C$ over the $n \times$ A/B/1 queuing model is achieved as a result of the *scaling effect* shown by Kleinrock (1974, 1976).

B-channel, while if the $n \times 64$ kbps method is used (see Chapter 6) several TSs will make up a superchannel. However, model 2 is similar to model 3 in its use of a single superchannel queue, hence resulting in a similar type of threshold control structure. This model will not be studied further for reasons of similarity with models 1 and 3.

5.2 THE CHANNEL MANAGEMENT ARCHITECTURE

The main requirements of the management architecture are the provision of efficient management functions, and practicability. The minimal set of channel management functions can be identified as:

- Ability to monitor and update the status of the interface under control
- Ability to set up and remove channels
- Ability to vary the bandwidth of logical channels

In all the architectural models described in the previous section, a further issue needs to be considered: that of whether *a priori* or negotiated superchannels are to be used. In the *a priori* method, all physical channels established between two communicating ends are assumed to be part of the same superchannel. This prevents the operation of multiple parallel superchannels across the same user–network interface and is less flexible. The negotiated superchannels method supports multiple superchannels (and ordinary channels) between two communicating ends. In model 1, the *a priori* method leads to a simpler interface status table structure and necessitates a different status block for reporting current interface status (see Sections 5.2.4 and

5.2.5). Both of these methods are dealt with in the following sections. The examples given will be mainly for the model 1—*a priori*—method, and differences indicated where necessary.

The main functional units necessary for channel management and the design of the messaging structures at the management–ISDN interface are described in the following sections.

5.2.1 The channel management functional units

The above requirements can simply be satisfied by the provision of the following *channel management functional units (CMFUs)*:

1. *Channel status table (CST)* Keeps the current state of the ISDN interface
2. *Channel management routine (CMR)* Interrogates the CST at specified epochs and makes a decision to set up or remove channels or vary their bandwidths according to some performance and/or cost criteria.

The channel status table is updated every time a packet is routed to the ISDN and whenever a message is received from the ISDN interface module reporting the completion of packet transmission and the UNI status. In its simpler form, the CST can be used to monitor only the outbound traffic. The two important metrics to be updated are the instantaneous *transmit queue size* and the *traffic arrival rate*. Complexity of the CST and its operation can be increased if incoming traffic is also monitored to record the incoming *traffic arrival rate* and the *receive queue size(s)*.

In the design of the architecture, placement of the main functional units can be implementation-specific. However, a convenient separation of the management operations at the *logical level* from those operations performed at the *physical interface level*, including any signalling protocol implementations at layers 1–3, eases the design effort and provides a much needed level of abstraction. Furthermore, there exists no standard software interface between the user applications and ISDN D-channel protocols at layer 3. Indeed, such a standardization may never be adopted. Therefore, such a software interface needs to be designed.

Two levels of decision processes are identified within the channel management function (routine): high-level and low-level.

High-level decision process (HLDP) At this level, a decision needs to be made to open/close a new/old logical channel to a new/existing destination. Once a logical channel is established, a decision is made to send or drop a packet, depending on the node/channel queue congestion level or other decision criteria. Furthermore, for dynamic control of channel bandwidth, a decision needs to be made according to some prescribed criteria (e.g. threshold control), and using a decision metric, whether to increase or decrease the channel bandwidth and by how much (Deniz and Knight 1989b). For this, a high-level primitive, *CHange_BandWidth (CHBW)*, could be used. Hence the following high-level primitives can be defined for use between the HLDP and lower-level modules:

1. *OPEN* (*isdn_address, bandwidth*) Low-level decision process (LLDP) module returns the assigned logical channel reference number (LCRN).
2. *CLOSE* (*LCRN*) Indicates the complete removal of all physical channels assigned to a logical channel (LC) with a given LCRN.

3. *SEND (LCRN, data_buffer)* LCRN value used to queue the packet to the correct logical channel queue.
4. *CHBW (LCRN, bandwidth, direction)* Indicates request for change in the bandwidth of an LC by the amount 'bandwidth'. Increase/decrease is indicated by the 'direction' value.

Of course, primitives are also needed to indicate the successful or non-successful completion of each command sent from the HLDP to the LLDP. These are OPENED, CLOSED, PKT_SENT and BW_CHANGED. The asynchronous type of interface suggested is necessary since the HLDP needs to be free to deal with other requests (e.g. SEND, PKT_SENT, etc.) while it is waiting for, say, OPENED for a previous OPEN command. This mode of operation is deemed to be better than, say, suspending all other operations while waiting for responses from the LLDP.

Flow charts of the routines are shown in Figs. 5.1 and 5.2. Here, the control metric is shown to be based on the channel queue length (*qlen*). The *check_qlen* ( ) function is used to determine whether any change in BW is necessitated as a result of threshold crossing.

Low level decision process (LLDP) The main duty of this decision process is to act upon the high-level primitives and interface with the B and D-channel processes. Furthermore, at this level a decision is made as to which types(s) of physical channel(s) (e.g. B or H-channel type) to use in order to aggregate the bandwidth to the requested logical channel bandwidth. The assignment of LCRNs and call references is also undertaken. A logical channel-to-physical channel(s) mapping is necessary at this point. A possible time slot mapping (TSMAP) table is shown in Fig. 5.3. A decision to close an existing connection and re-establish the connection using different types of physical channel structures will also need to be placed here if such a functionality is supported. In the case where H-channels are used, another decision is needed to determine the TS(s) to be assigned to each physical channel. This may be determined by the TS hunting strategies adopted. However, the dynamic reassignment of TSs to physical channels for an existing call is not supported in the current PRISDN UNI. In the case where only B-channels exist, CCITT Rec. I.450/1 specifies that each TS number also corresponds to the channel number.

5.2.2 Placing the CMFUs

As mentioned in the previous section, this is very much dependent upon the hardware and software configurations as well as on performance issues. A broad assumption that the network layer relay is placed on a hardware configuration where a main processor board controls the access to each network type, each of which is interfaced by an interface board (see Appendix I) with on-board interface software for low-level protocol/hardware interfacing, is not unrealistic, as there are many examples of this in the existing networks. One such hardware set-up is given in Appendix I. According to the above assumptions, two broad classifications can be made with regard to the ISDN interface. In the first, the ISDN hardware interface could be composed of a single board (type 1) or multiple boards (type 2). The second class of interface may be necessitated for speed reasons where each interface board is given the responsibility for a limited number of B-channels as well as separating the B and D-channel boards. An example of this is the PROOF LAN–ISDN relay (Knight 1989), where handling of X.25 protocols over the B-channels is also required from the interface boards.

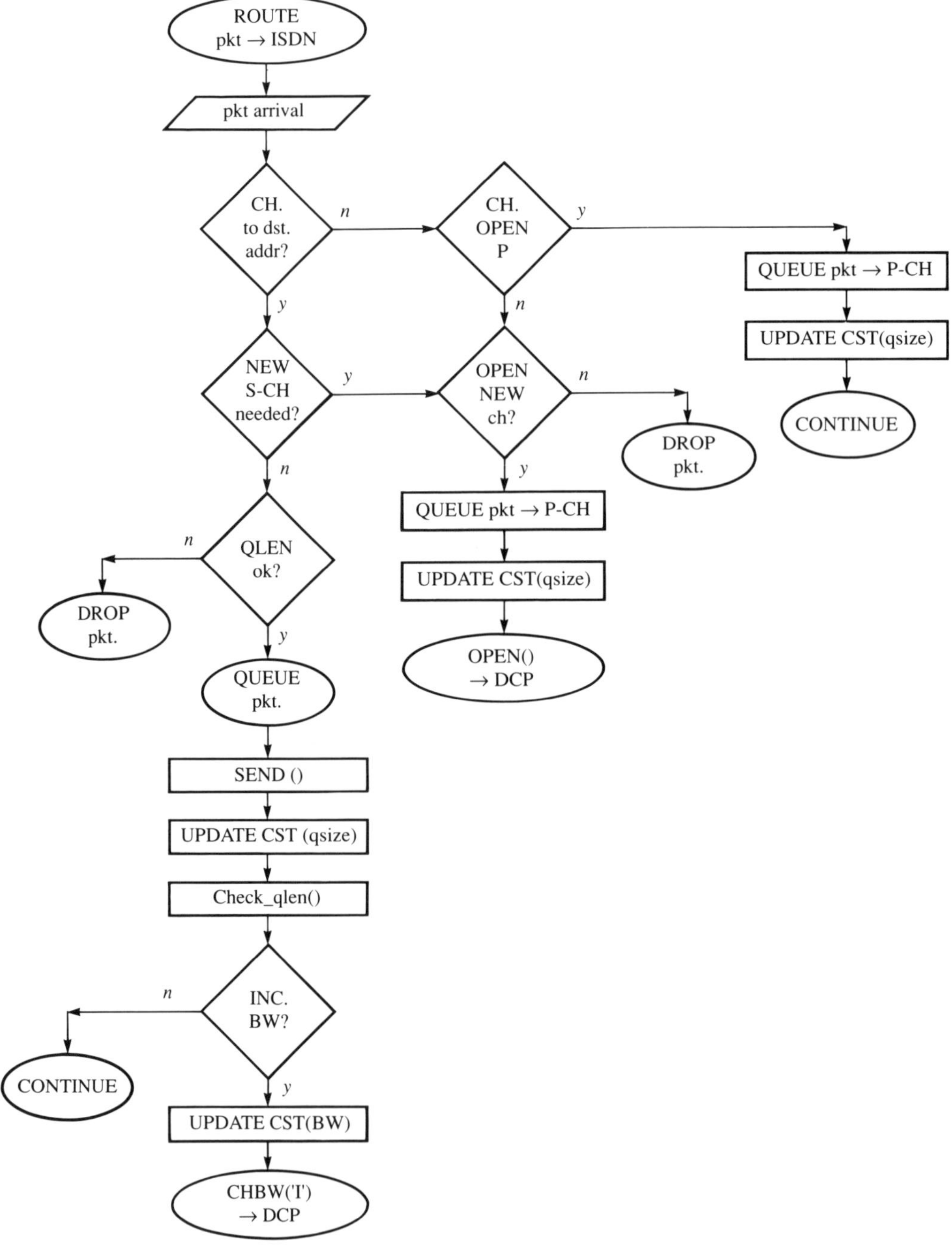

pkt: packet; CH (ch.): channel; dst: destination; addr: address; P: pending; P-CH: pending channel
qsize: queue size; BW: ISDN bandwidth; INC: increment; CST: channel status table; DCP: D-channel process
CHBW: change bandwidth function

Figure 5.1 High-level decision process: action routine

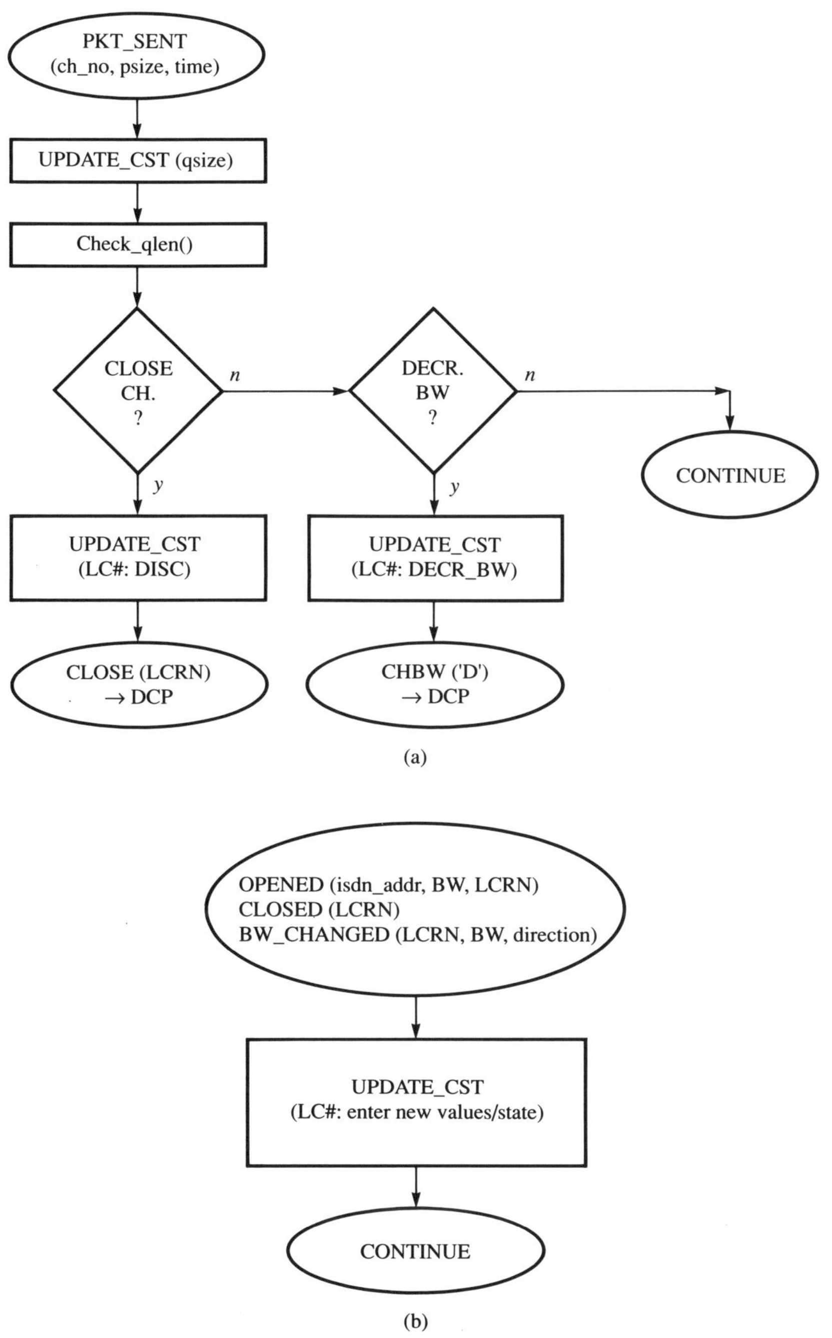

LC: logical channel; LCRN: logical channel number; DISC: disconnect
DECR_BW: decrement bandwidth; psize: packet size

Figure 5.2 High-level decision process: reply routines

TS	PCRN (Call_Ref)	LCRN
0	•	•
1	25	121
•	•	•
31	57	121

PCRN: physical channel reference number; LCRN: logical channel reference number

Figure 5.3 Details of a TSMAP table

Thus, the channel management functional units (CMFUs) can be placed either within the *main processor* or the *interface processor* domains. In the type 1 interface it may be simpler to place the CMFUs in the interface processor domain; this will avoid excessive inter-processor messaging between the main processor processes and the interface processes over the system bus. In fact, even the 'routing' function[1] can be placed within the interface processor domain. In the type 2 interface, however, placement of the CMFUs at the main processor domain may simplify the inter-board messaging by combining the decision process at the main board upon which the correct interface board is activated. In the rest of this chapter, a model is described whereby the CMFUs are divided between the main and interface processor domains.

Figure 5.4 shows the reference configuration for a channel management architecture. Here, the high-level decision process (HLDP) is left within the channel management routine (main processor domain) while the low-level decision process (LLDP) is placed within the interface processor domain. This would minimize the number of message exchanges and the speed of response since, once the LLDP is called, it will result in either a D-channel protocol activation (OPEN, CLOSE, CHBW primitives) or a B-channel activity (SEND primitive). The LLDP is shared between the B and D-channel processes of the interface processor.

5.2.3 Designing the channel control messaging interface

From the reference configuration shown in Fig. 5.4, it can be seen that the B and D-channel interfaces can be addressed in a unified manner. This approach has been adopted in the design of the channel control messaging structure. A Channel Control message Block (CCB) is designed to achieve the exchange of information between the high and low-level decision processes. The structure of these message blocks is similar to the Q.931 message structure (CCITT 1988a) and is shown in Fig. 5.5.

5.2.4 The channel status table (CST)

The channel status table contains all the information about the ISDN interface necessary for decision-making by the CMR. The form and structure of this table could be implementation-specific. Also, depending on whether the model 1 or model 3 architecture type is used, its information content may be different. This will also affect the form of the status block used to interrogate the CST.

[1] In its simplest form, this is an address resolution function and applies to the point-to-point type links we are considering.

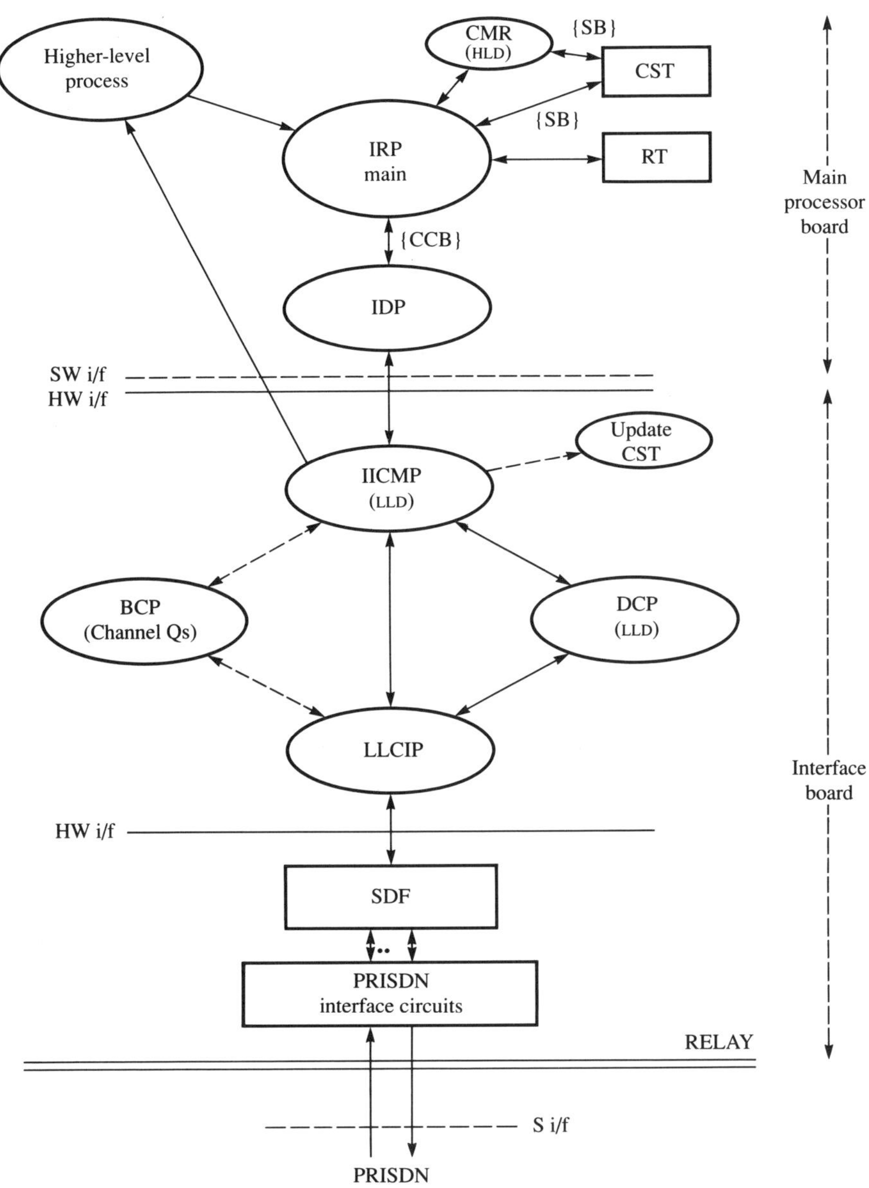

CMR: channel management routine (within IRP); HLD: high-level decision
SB: status block; CST: channel status table; RT: routing table
CCB: channel control block; IDP: ISDN port driver process
IICMP: ISDN interface control main process; BCP: B-channel process
DCP: D-channel process; LLCIP: Low-level control interface process
LLD: Low-level decision; SDF: synchronous data formatter (HDLC)

Figure 5.4 Reference configuration for CMFU placement

Protocol discriminator
Call reference
Message type
Information element (Mandatory/optional)

(a)

Call_Ref
Msg_Type
Ch_Type
Ch_No/LCRN
Issrc_Add
Isdst_Add
Cause
Bandwidth

(b)

Figure 5.5 Details of (a) Q.931 message structure; (b) channel control block (CCB)

A sample CST for model 1 is shown in Fig. 5.6(a). Here, the state of each channel (TS) can be monitored. It is assumed that the *a priori* type superchannel formation is used. This is significant in that a search of the table for a given ISDN destination address (*isdst_add*) will reveal all the B-channels connected to that destination. A packet can then be scheduled to the most appropriate queue according to the scheduling policy being used (e.g. shortest queue). The status of those channels connected to the destination is returned by the status block (see Fig. 5.7). A sample CST for model 3 is shown in Fig. 5.6(b).

Ch no.	Call ref.	Ch. state	BW	Tx pkts	Tx Ql	Act	QoS	Call src	Time out	ipsrc add	ipdst add	isdst add
1												
.												
30												

(a)

SC no.	LCRN	SC state	BW	Tx pkts	Tx Ql	Dst LCRN	Act	QoS	Time out	ipsrc add	ipdst add	isdst add
1												
.												
30												

(b)

Ch: channel; Ref: reference; BW: bandwidth; Tx: transmit; Pkts: packets; Act: activity (traffic rate)
Ql: queue length; src: source; dst: destination; ip: Internet protocol; isdst: ISDN destination
add: address; SC: superchannel; LCRN: logical channel reference number

Figure 5.6 Details of the channel status tables: (a) model 1 (b) model 3

5.2.5 The channel management routine (CMR)

This routine can interrogate the CST by a messaging block called the *status message block* (*SB*) and a function called, *check_con (isdn_address, sb_p)*, where 'check_con' is short for *check_connection* and *sb_p* is a pointer to the status message block.

After an interrogation, the SB contains the necessary information for the CMR to make a decision described in the HLDP, and, using one of the described primitives, it activates the LLDP for the correct action across the main processor–port interface. Figure 5.7 shows the SB structure and fields applicable to model 1 with *a priori* superchannel formation.

	Ch no.	Call ref.	Ch. sta.	Qued pkts	Qlen
1					
•					
m					

	Full	No. of cons.	No. of free chs.	Free ch. no.	No. of resv. chs.	Resv. ch. no.	ISDN dst. add	Send ch. no.	C1	C2
1										

Sta: state; Cons: connections; Full: channels (not) busy; Resv: reserved; C1,C2: code fields

Figure 5.7 Details of the status message block (SB)

5.2.6 Status updating

As a result of each instance of interface and channel activity, the CST is updated by a procedure called *update_cst (io_blk_p, sb_p, msg)*, where *io_blk_p* is the pointer to an input–output message block containing the data packet, *sb_p* is the pointer to the status block (SB) and *msg* is the message indicating the type of operation causing the update. The message types are broadly classified as those pertaining to the D-channel activity and to the B-channel activity. The first type affects mainly the channel state, while the second type affects mainly the updating of B-channel queues and activity rate fields. The messages used for the D-channel activity are: SETUP, RESV, CONN, CONN_ACK, DISC, REL, FREE and UPDATE_BW. These messages correspond to the D-channel protocol messages/states at layer 3, i.e. Setup, Reserve, Connect, Connect_Acknowledge, Disconnect, Release and Free. The message UPDATE_BW results from a successful CHBW( ) functional call to the interface. A further message for Release Complete can also be implemented. (See also Appendix III for simulator implementation.) The messages used for the B-channel activity are UPDATE_QLEN, UPDATE_ACT and UPDATE_TO. Every time a packet is sent to the interface for channel queues and when packets are transmitted from the interface, the UPDATE_QLEN message is used to update the queue size and number of packets in the system. The activity message UPDATE_ACT is used to measure the packet arrival rate (over a period) or the last time a channel was used. The UPDATE_TO message can be used to record/modify the time-out values set for each channel.

5.3 A DCMA FOR ETHERNET–ISDN RELAY

The main principles of the dynamic channel management architecture discussed in the previous section can be applied to the design of a DCMA for the Ethernet–ISDN relay. Assuming a rather simple-minded connectionless network layer relay, all of the DCMA functionalities described can be implemented. Indeed, it is this design that has been used in the implementation of our LAN–ISDN relay/ISDN simulator (ISIM) described in Appendices I–III. Furthermore, the ISIM configured for model 3 is used in the bandwidth management policy performance measurements described in Chapters 7 and 8.

In the design of the DCMA, several assumptions have been made regarding the capabilities of the ISDN interface hardware (see also Appendix I). These assumptions are based mainly on the VLSI chips available today or soon to be available. It is assumed that the ISDN interface board has enough processing power derived from a single or multiple processors to manage a PRISDN interface, including all the B-channels as well as the signalling D-channel. The existence of enough buffer space for the transmit and receive buffers for each channel is also assumed. It is further assumed that a form of framing will be present or established to implement the data link functions on the B-channels. The main processor can communicate with the interface processor(s) using global memory and passing pointers to data/message structures. A common interface for signalling and data paths between the ISDN router process (IRP) and the ISDN interface modules is designed in line with Figs 5.4 and 5.9. This interface is assumed to be capable of specifying connection set-up or removal on multiple B-channels at the same time.

5.3.1 Protocols

The relay reference protocol structure is shown in Fig. 5.8. This protocol structure has previously been suggested in Deniz and Knight (1989b). The CL network protocol (CLNP) could be replaced by the internet protocol (IP) in networking environments using the TCP/IP suite (DARPA transport control protocol/internet protocol). The data link layer protocol on the Ethernet could be null in some implementations, in which case only the MAC frames would be used. As the same network layer protocols are used over both the Ethernet and ISDN, the CLNP relaying is rather simple; it involves the extraction of the CLNP packet from one link layer, framing and placing it in the other network's framing. Of course, issues such as address resolution, fragmentation and re-assembly need to be considered. In the case of the circuit-switched ISDN, the existence of a channel to a given destination must also be verified. A channel management routine within the relaying function acts like a decision process, determining whether a new circuit should be opened, an existing one closed, the bandwidth of an existing channel changed or a network layer (NL) packet simply forwarded over an existing channel. The NL packets are then relayed over the ISDN B-channel(s) within core HDLC frames assuming that a channel to the given destination exists. For channel management simulations, only the basic D-channel signalling functions of I.451 (layer 3) and LAPD (layer 2) need to be implemented in a much simplified manner. (For details of ISDN D-channel protocols, see Chapter 2.)

	RELAYING FUNCTION		
Network layer	CLNP	CLNP	I.451
Link layer	LLC1	HDLC	LAPD
Physical layer	CSMA/CD	I.431	I.431
		B	D
	Ethernet	ISDN	

CLNP: connectionless network protocol; LLCl: connectionless logical link protocol
LAPD: link access protocol for the D-channel; HDLC: HDLC framing
B, D: ISDN user and signalling channels, respectively

Figure 5.8 CL Ethernet/ISDN relay protocols

5.3.2 Main processes

The main tasks within the Ethernet/ISDN relay (EIR) are shown in Fig. 5.9. The IP relaying process (IPRP) performs all the general relaying functions including buffer management, initial address resolution, fragmentation and reassembly. It calls the address resolution protocol (ARP) or the ISDN router process (IRP) for address resolution functions to Ethernet and ISDN, respectively. The ISDN router process achieves the IP–ISDN address mapping via a fixed routing table and monitors the status of ISDN interface via a channel status table (CST) shown in Fig. 5.6. The IRP may not include a full routing function if only point-to-point communication is used or if only one interface type exists between the Ethernet and ISDN. The call control and message forwarding for each channel buffer is accomplished by the use of a channel control message block (CCB) used between the ISDN router (IRP) and driver (IDP) modules (see Fig. 5.10). The status information regarding the PRISDN interface board(s) is returned using the same CCB message structure and is used for CST updating. The CST status can be called up at any time by the IRP or the channel management routine, which itself is activated by the IRP every time a datagram arrives at the relay and is to be routed to the ISDN output interface. CMR is also activated when a packet transmission occurs and its completion is reported back by the ISDN interface. CST status is reported back using a status message block (SB). Figure 5.7 shows the SB, while Fig. 5.10 shows a detailed diagram of the IRP and IDP processes of the EIR.

In Fig. 5.9, all the processes running on the relay main processor are shown within the relay module (RLM). Two further modules, the user module (UM) and the ISDN interface module (IIM), are defined; these are used in the simulation implementation and are referred to in Appendix I for simulation modelling. Note that these arbitrary module names are for simulation purposes only. The CCB message structure is used for communicating with both the B and D-channel processes in the IIM. This represents a unified and simpler approach to ISDN interface access. With this approach, the actual interface hardware implementation could be assumed to be on single or multiple boards using single or multiple processors. Within the ISDN interface module, a circuit control status table (CCST) can be used optionally to monitor and manage the status of individual ISDN channels (similar to the CST)

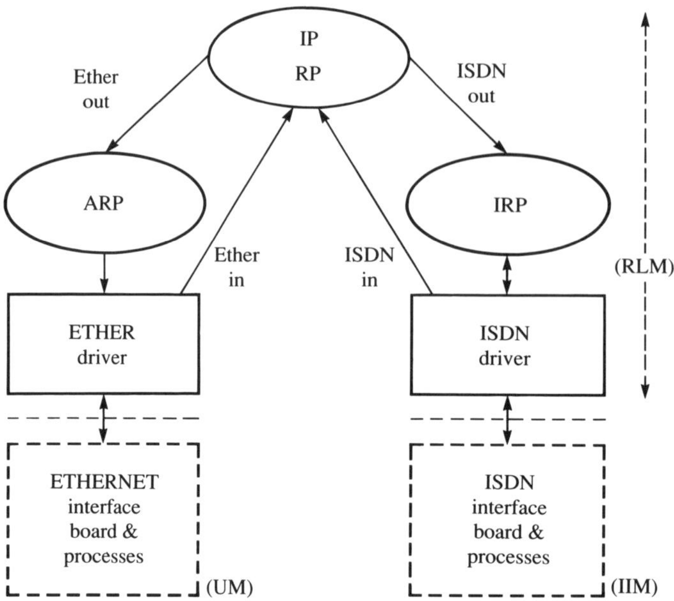

ARP: address resolution protocol; IPRP: internet protocol relaying process; UM: user module
IRP: ISDN router process; IIM: ISDN interface module; RLM: relay module

Figure 5.9 Ethernet/ISDN relay processes

as well as to transmit and receive channel queues (CQs). This depends on whether or not the CST data structures are globally shared. If they are shared, as shown in Fig. 5.4, a common CST updating function can be used by both the main and interface processors. Within the EIR core functions the same message blocks are used, and their pointers are passed to processes along the way. This ensures that there is as little buffer copying as possible for operational speed. Communications between the core functions and the driver modules are established using global memory. The same is assumed between the driver module and the PRISDN interface board(s).

5.4 SUMMARY

This chapter has presented a dynamic channel management architecture design based on the status table approach. The decision processes involved in the management function are classified into the high-level and low-level processes. It is shown that a high-level decision process using simple primitives can implement opening, closing and bandwidth variation of channels as well as communicate packets to the transmission interface. The main channel management functional units (CMFUs) are identified as the channel status table and the channel management routine. The importance of the placement of CMFUs is discussed with reference to different interface configurations. It is concluded that the placement of CMFUs is dependent upon the hardware and software configurations as well as nodal performance. A

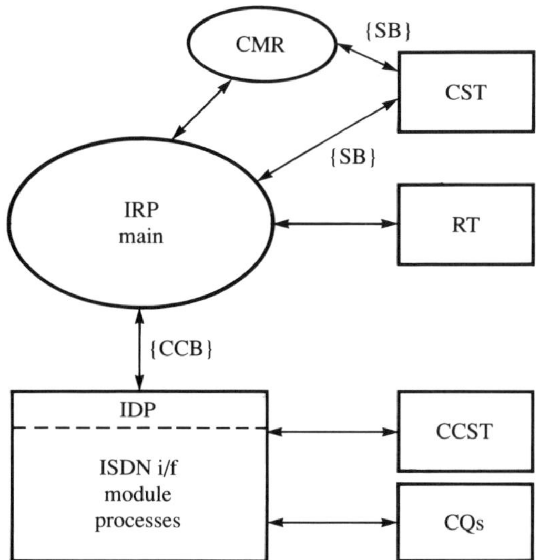

CMR: channel management routine; CST: channel status table
RT: routing table; SB: status block; CCB: channel control block
CCST: circuit control status table; CQs: channel queues; IDP: ISDN (interface) driver process

Figure 5.10 Details of the ISDN relay process (IRP)

simple channel control messaging interface is designed and shown to be effective in combining D and B-channel messaging. Finally, a DCMA for Ethernet–ISDN relay is designed and its operation described. This design is used in the simulation model developed for performance measurements of different bandwidth management policies in the following chapters.

CHAPTER

SIX

SUPERCHANNELS AND THEIR PROTOCOLS

In ISDN, user channels can be established, maintained and released by using the D-channel protocol (CCITT 1988a). For these basic operations some form of channel/call state information must be kept by each user–ISDN interface. Such a state information, combined with control metrics, can form a powerful method of dynamic channel management scheme. A management architecture for such a scheme within a connectionless network layer relay at the LAN–ISDN interconnection, based on the status table approach, has been proposed in Chapter 5 and in Deniz and Knight (1989a, b). Here, the problems associated with superchannel formation are analysed and relevant solutions are proposed. Although some of the arguments presented in this chapter apply equally to both the packet- and circuit-switched services in ISDN, the discussion is limited mainly to the circuit-switched ISDN service.

Section 6.1 provides a background to the problems encountered in superchannel formation and dynamic bandwidth variation. Section 6.2 defines some terms and introduces some concepts that are used in the rest of the chapter. The scope of the discussions in the chapter is also described. The time slot sequence integrity (TSSI) problem is discussed briefly in Section 6.3. Section 6.4 presents a short review of the facilities for call and channel identification in an ISDN interface using the layer 3 signalling protocols for the D-channel, I.451 (Q.931) (CCITT 1988a). The *a priori* superchannel formation is also defined. In Section 6.5, protocol issues pertaining to the use and identification of superchannels and various relevant reference configurations are examined; in addition, the usefulness of the user–user signalling facility in the ISDNs is investigated. In Section 6.5.2, a novel protocol which can be used over the user–user signalling for superchannel identification is proposed and its operation described. In Sections 6.5.3 and 6.5.4, various modifications applicable to the channel identification and call reference information elements of the Q.931 signalling protocol for end-to-end superchannel identification are described. The dynamic bandwidth variation problems and proposed solutions for two types of superchannel formation are considered in Section 6.6. Finally, Section 6.7 presents the main conclusions and a summary of the chapter.

6.1 BACKGROUND

Current CCITT ISDN standards (CCITT 1988a) specify fixed-sized physical channels (B and three H-channel sizes—H_0, H_{11}, H_{12}—corresponding to 1, 6, 24 and 30 time slots (TSs) in a TDM frame) at the user–network interface (UNI). In PRISDN access at 2.048 Mbps configured as 30 B-channels, each B-channel corresponds to a TS and a transmission capacity of 64 kbps, while two time slots are reserved for channel control and synchronization information. Channel sizes of arbitrary integral TSs of size $n \times 64$ kbps are yet not available although they are desirable when traffic includes services such as LAN–LAN interconnection, compressed video or multi-media conference. Such channels are termed 'superchannels' in this book. Superchannel formation can be considered under two distinct cases:

1. Formation by aggregation of fixed-size channels (e.g. nB + mH) outside the network.
2. Formation by the $n \times 64$ kbps method by the network itself.

Here, a distinction between the $n \times$ B and $n \times 64$ kbps service is made such that the former needs n separate B-channel connections, while the latter can be requested in one connection set-up. When case 2 service is made available, the need for case 1 will be obviated. However, the ISDN will provide only fixed-size channels for the immediate future.

Four fundamental problems are identified with the formation of superchannels:

1. *Provision of time slot sequence integrity (TSSI) across the ISDN* Loss of TSSI can occur as a result of different time delays encountered by the TSs comprising a superchannel, causing skew between TSs and rearrangement within a frame of TSs arising from TS switching within ISDN. This can occur within a digital network employing TDM, because each TS could be switched through the network via different trunk lines and switching nodes. The TSSI provision is difficult because switching a bundle of TSs through the same TDM trunks via the network increases the call blocking probability. In the case of type 1 superchannel formation, no guarantee can be given for using the same switching nodes and paths for different physical channels. In type 2 superchannel formation, the network may or may not provide the TSSI.
2. *Lack of provision for the identification and management of type 1 superchannels between the end users* (note that the network may not need to know anything about this association). For example, if two B-channel connections are established by separate call set-ups between two end-users, the called party must be aware that these two channels are to be used as a superchannel (i.e. transferring consecutive octets of a packet alternately). An *a priori* agreement for aggregating all existing and new channels between two destinations may not be advantageous. Sometimes it may be desirable to maintain several channels to the same destination so that different classes of traffic can be split between them.
3. *Dynamic variation of superchannel bandwidth* In the case where a superchannel is formed by the aggregation of different channel sizes, the reduction of channel bandwidth may necessitate the removal of an existing wider channel connection and the establishment of multiple smaller-size channel connections.
4. *Lack of D-channel protocol facility* to vary the bandwidth of a superchannel dynamically when type 2 superchannel formation method is used.

In this book, it is assumed that some form of corrective measures are taken to ensure TSSI between end-points in a communications link. Several viable proposals for TSSI provision in multi-slot channel ISDN interfaces already exist (Altarah and Motard 1989; Boltz *et al.* 1989; Burren 1989). These measures can be implemented in hardware or software; they could be undertaken by the network, the network terminator, customer premises equipment (CPE) or the customer's equipment. If carried out outside the customer's equipment (e.g. by a black box or CPE), some form of superchannel identification is necessary between the customer's equipment and the TSSI provider.

For the superchannel identification problem, in the case of superchannels formed by the aggregation of fixed-size channels, the usage of either the optional user–user (signalling) information element within a SETUP message or an extension to the channel identification information element or a modification to the call reference information element uniquely to identify a superchannel is proposed. If the superchannels are formed by the $n \times 64$ kbps service, no identification problem exists since the network itself must provide it. Indeed, the layer 3 D-channel protocol (Q.931) has a built-in facility (see Fig. 6.6) for handling multiple time slot channel requests. However, the use of this facility is currently limited to the CCITT-specified fixed channel sizes as already described. For the dynamic bandwidth variation of superchannels, modifications to the Q.931 protocol are proposed in this chapter and in Deniz and Knight (1990).

6.2 DEFINITIONS

The following definitions are assumed in what follows:

- *Basic bandwidth unit* A basic bandwidth unit (BBU) is the smallest amount of bandwidth that can be switched as a complete entity. Current CCITT recommendations specify a BBU of 64 kbps which corresponds to a single time slot (TS) in successive TDM frames.
- *Physical channel* This is a channel provided by the network which is composed of one or more BBUs (or TSs). A physical channel is also switchable as a complete entity. Current CCITT standards specify two types of physical channel: B and H (sub-classified as H_0, H_{11} and H_{12}). The physical channel reference number (PCRN) is the call reference number (CRN) assigned at the time of SETUP since each call can be made only at the allowed physical channel capacity units. Also, the bandwidth of a physical channel cannot be varied dynamically. Furthermore, it is composed of *single connections*, where all of its bandwidth units are assigned in one call set-up. In the rest of this book, the terms PCRN and the CRN are used interchangeably.
- *Logical channel* This is a channel of 'arbitrary' bandwidth (multi-slot) composed of one or more physical channels. When composed of only one physical channel, its logical channel reference number (LCRN) has a one-to-one mapping to the call reference number of a physical channel. The bandwidth size of a logical channel may be dynamically varied. Also, it may be composed of *multiple connections* to the same destination. Figure 6.1 shows the relationship between the physical and superchannels.

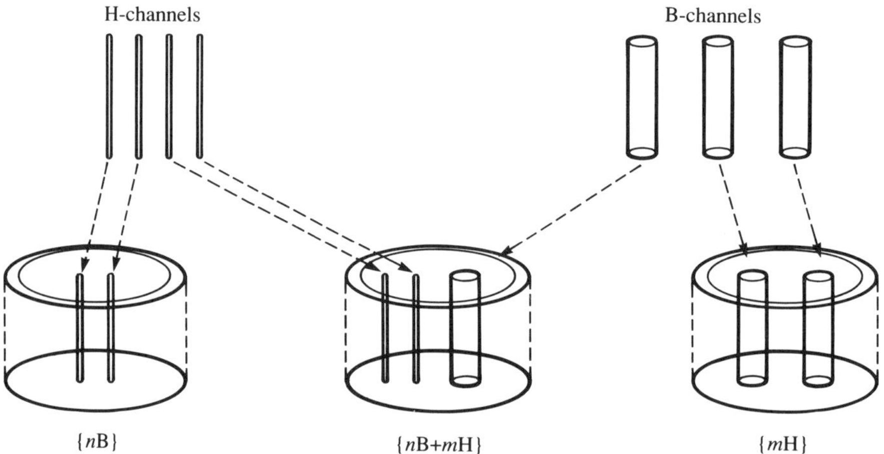

Figure 6.1 Formation of superchannels using physical channels in ISDNs

The scope of this chapter includes:

1 *Dynamic allocation* The bandwidth (total bit rate) assigned to a logical channel can be varied dynamically.
2. *Multiple connections* A logical channel may be formed by multiple connections of physical channels. In this case, an LCRN is mapped to multiple PCRNs since each connection can be identified by its unique CRN.

The following differences are assumed to exist between $n \times$ B and $n \times$ 64 kbps services:

- *n X B service* This is a superchannel formed by the aggregation of n separate B-channel connections. If a black box or CPE provides the TSSI, a protocol interface capable of requesting all n channels within one call set-up request and dynamic bandwidth variation facility must be built between the user and the black box.
- *n X 64 kbps service* This is a service provided by the network which may (or may not) provide the TSSI. If the network does not provide TSSI, the black box solution of $n \times B$ service can be applied. However, in the rest of this book it is assumed that the network also provides TSSI. Furthermore, all $n \times$ 64 kbps units can be requested in one call set-up, and dynamic bandwidth variation is allowed. For this, a protocol facility is suggested in the latter sections.

6.3 THE TIME SLOT SEQUENCE INTEGRITY PROBLEM

The CCITT Rec. I.340 on the ISDN connection types (CCITT 1988a) defines the multi-slot H-channels (see Section 6.1) as 'unrestricted' in their information transfer capability and '8 kHz structural integrity' in their structure attributes. These properties imply that the network will maintain time slot sequence integrity (TSSI) for these channel types. It is also noted that some networks will not support these connection types until some future date, and as yet there are no Recommendations available for the switching of H_0 and H_1 channels; the current switching fabric is based on 64 kbps switches. Furthermore, the definition of other H-channels is left for further study.

In general, the provision of TSSI within a digital network for multi-slot channels is not automatic. This is because the individual 64 kbps constituent channels could follow different physical paths and could be variably delayed at each switching centre en route. The total effect is the experiencing of different propagation delays through the network. Two different approaches can be taken in the provision of TSSI for $n \times 64$ kbps services: within the network, or within the customer premises. Figure 6.2 shows the schematic for TSSI provision domains.

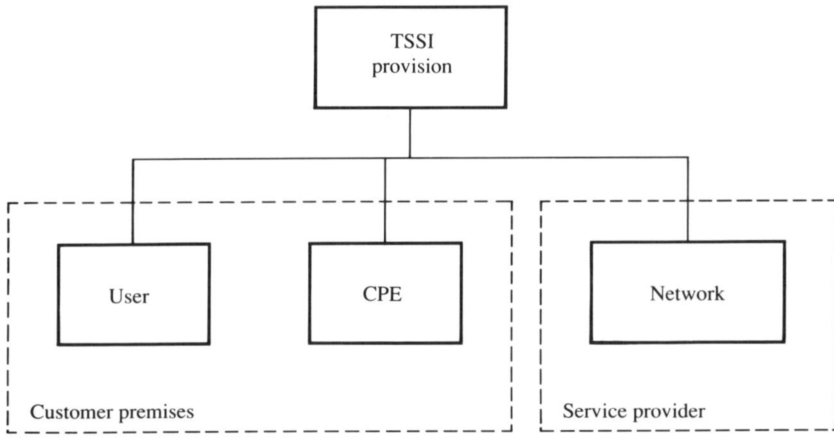

TSSI: time slot sequence integrity; CPE: customer premises equipment

Figure 6.2 TSSI Provision Domains

6.3.1 Provision of TSSI within the network

For an $n \times 64$ kbps connection, the TSSI can be maintained by guaranteeing the switching of all n channels through the same TDM system on each network link, hence ensuring equal transmission times. Also, paths through switching networks must be chosen so that the time slot sequence is maintained. However, implementation of such a path selection strategy requires modifications to existing exchange software (and possibly hardware). The need to place all n channels in the same TDM system is a severe constraint on the choice of path and leads to excessive blocking probabilities (BPs) for such calls (Boltz *et al.* 1989). It is reported that these BPs become unacceptable for calls requiring more than three channels (Altarah and Motard 1989).

A second solution to this problem is to build exchanges that can switch channel entities at multi-slot bit rates (e.g. H_0 and H_1). All the above solutions constitute major changes to the existing integrated digital network (see Boltz *et al.* 1989) and still do not allow dynamic variation of a multi-slot connection bandwidth. This is because the new 64 kbps channel(s) to be added to the existing multi-slot connection may not be guaranteed to be switched through the same nodes and TDM trunks in the network.

An alternative is to have each of the destination local exchanges provide TSSI using some form of TS delay equalization based on a training sequence, measuring the relative delays encountered by the constituent TSs of a superchannel as they travel through the network. This would also cater for the problems encountered when the bandwidth of a superchannel is varied dynamically during the lifetime of a connection. However, this seems to be a rather unpopular solution for the service providers as it necessitates extensive changes to their local exchanges.

6.3.2 Provision of TSSI within customer premises

This can be done in two ways: within the user equipment (UE), or within a customer premises equipment (CPE). Various methods have been proposed by Altarah and Motard (1989), Boltz *et al.* (1989), Burren (1989) and others (e.g. Australia 1991) for the provision of TSSI and $n \times 64$ kbps service. All of these proposals provide some methods for the synchronization of the TS at the receiving end with those at the sending end so that the time slot sequence is maintained. None of the above studies looks at the problem of superchannel formation using the mixture of B and H-channels; indeed, only the draft Australian Standard (Australia 1991) deals with the problems of superchannel identification. Moreover, none deals with the dynamic bandwidth variation of such channels from the CCITT signalling protocol aspect. These problems are addressed in the rest of this chapter. In the following chapter the proposed solutions to the TSSI problem are examined in greater detail and the draft Australian Standard's proposal for the superchannel identification is presented.

6.4 CALL AND CHANNEL IDENTIFICATION

A major problem in the superchannel formation is the unique identification of superchannels at each end of an ISDN connection. Three possible methods for superchannel identification are identified. The first involves the design and use of a user protocol for superchannel identification at the UNI and is transparent to the network; it involves no modifications to the existing Q.931 protocol and utilizes the user–user signalling feature of ISDNs. The second and third methods rely on a modification of the Q.931 information elements for channel identification and/or call reference. Either could be used for informing the network of a superchannel association at a local interface. However, both of these information elements have significance only at the local UNI. For these methods to work, the co-operation of the network is needed in transferring this information to the destination interface and giving local significance to a similar set of assigned values (similar to the operation of the current call reference value). These are described in detail in Section 6.5.

6.4.1 The channel structures and time slot mappings

The CCITT Recommendations leave the TS-to-channel assignment at the PRISDN user–network interface composed of a mixture of B and H-channels to the user. Hence, in order to create an interface with superchannel facility, a TS-to-physical channel mapping as well as TS and/or physical channel-to-logical channel mapping are needed. Note also that, for the PRISDN interface that comprises B-channels only, the TS number in a frame also corresponds to its physical channel number. The general TS-to-physical channel assignment for larger physical channel structures could be based on a scheme that simplifies the associated channel hunting strategies. PRISDN interface chips are now available which are capable of assigning any number of TSs to any channel and hence of determining the channel bandwidths dynamically. For such an operation, a physical channel-to-TS(s) binding is necessary. This could be done dynamically within the user equipment. The existence of such a functionality is assumed.

6.4.2 Call and channel identification in PRISDN

According to Q.931 (CCITT 1988a), each call across the UNI is identified by a call reference number (CRN). This is a mandatory information element in the SETUP message as well as many other call control messages. Among other purposes, the channel identification information element (CIIE) is used for the TS-to-physical channel mappings across the UNI.

6.4.3 *A priori* superchannel formation

Superchannel formation by an *a priori* agreement for automatically aggregating all physical channels established between two co-operating ISDN sites is another solution to the superchannel identification problem. However, this may not be desirable if different channels are required to be set-up between two sites for applications with widely differing quality-of-service (QoS) parameters and traffic types (e.g. pulse code modulation (PCM) voice and packet data). For this reason, the *a priori* superchannel formation method is not assumed in this chapter.

6.5 PROTOCOL ISSUES

Several options exist for interfacing the TSSI providers described in the previous sections. These can be classified as follows, depending upon the distribution of information and TSSI provision functionality:

- *Case 1* The network (ISDN) does not know of the superchannel association.
- *Case 2* The network knows about the superchannel association and provides end-to-end assistance.
- *Case 3* The network provides the superchannels.

In case 1, the provision of TSSI is left to the user or CPE. Hence the superchannel identification and dynamic bandwidth variation are made transparent to the network between co-operating users. Similarly, in case 2 the provision of TSSI is made at the customer premises although the network helps in the identification of superchannels on an end-to-end basis. In case 3, the network provides the TSSI and there is no superchannel identification problem since the physical channel now corresponds to the logical channel. However, provision of dynamic bandwidth variation needs to be addressed.

Some possible solutions that would pertain to the above three cases are:

1. *Use of supplementary services within the existing Q.931 protocol* The user–user signalling feature of the ISDN can be used to inform the end-users (or CPEs) of a superchannel association, namely of the physical channels 'belonging' to a superchannel.
2. *Use of modified Q.931 (Q.931+) for superchannel identification* The call reference or the channel identification information elements could be modified to include fields for superchannel identification.
3. *Use of enhanced Q.931 (Q.931++) for dynamic bandwidth variation* This necessitates either the definition of a new message type (Vary Bandwidth) or the reuse of the SETUP message (see the following sections).

Various reference models applicable to the above cases are shown in Fig. 6.3. The relationship between the solutions and their protocol implications is shown in Fig. 6.4.

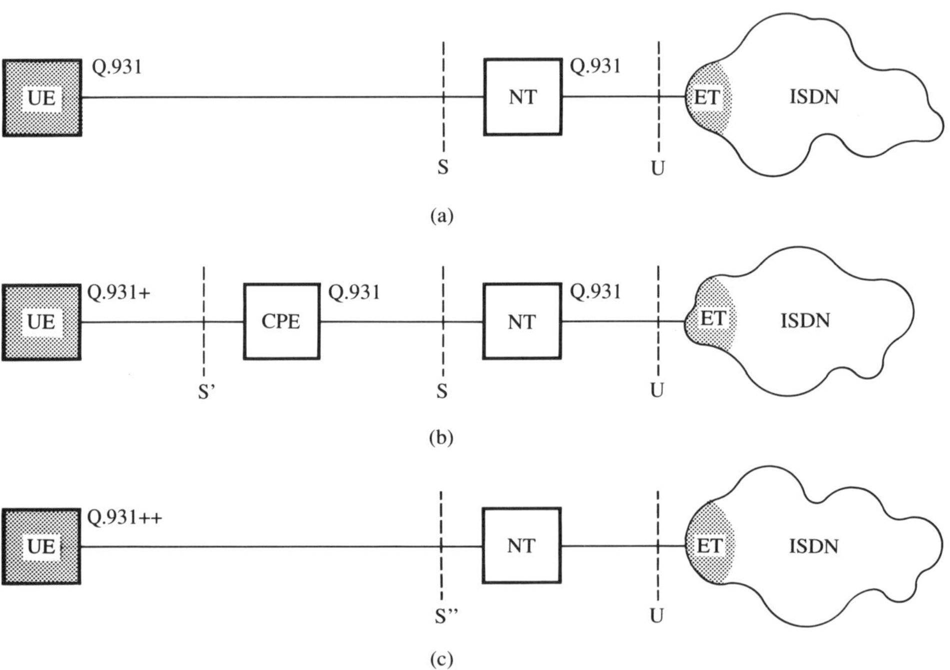

NT: network termination; ET: exchange termination
UE: user equipment; CPE: customer premises equipment
S: the ISDN 'S' interface; S' and S": enhanced 'S' interfaces; U: the ISDN 'U' interface

Figure 6.3 Various reference models for superchannel formation: (a) TSSI provided by user; (b) TSSI provided by CPE; (c) TSSI and variable bandwidth provided by network

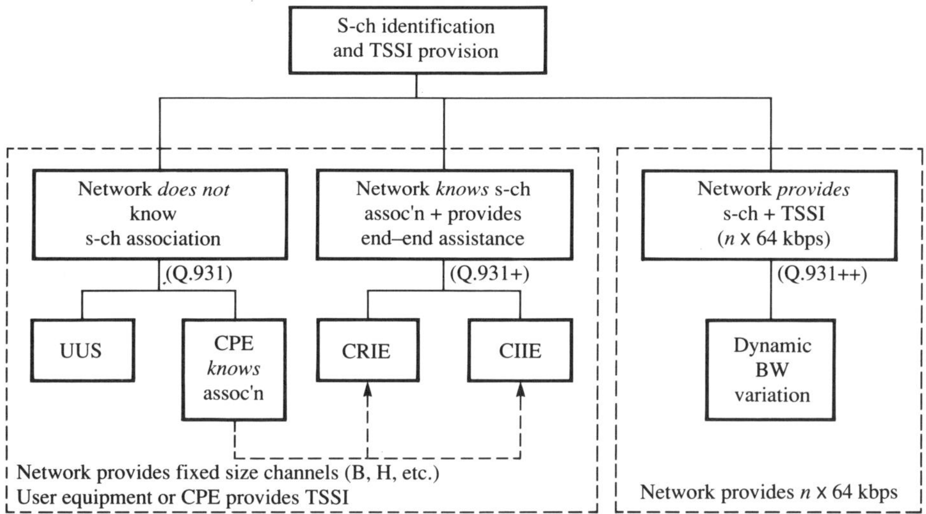

S-ch: superchannel; UUS: user–user signalling; CPE: customer premises equipment
CRIE: call reference information element; CIIE: channel identification information element
TSSI: time slot sequence integrity

Figure 6.4 Protocol implications for superchannel interfacing

6.5.1 User–user signalling (UUS)

This is the easiest method to implement, since it involves no changes to the existing protocols and no action by the network provider (apart from supporting the UUS). The UUS is transparent to the network and there are no restrictions on its contents. There are two broad classes of UUS within the ISDN as defined in CCITT (1988a). These are the UUS associated and the UUS not associated with circuit-switched (CS) calls. In the UUS associated with a CS call, there are three types of services:

- *Service 1* U–U signalling exchanged during the set-up and clearing phases of a call, within Q.931 call control messages (using the U–U information element — UUIE).
- *Service 2* U–U signalling exchanged during call establishment, between the ALERTING and CONNECT messages, within USER–INFORMATION messages.
- *Service 3* U–U signalling exchanged while a call is in the ACTIVE state, within USER–INFORMATION messages.

In the UUS not associated with a CS call, a temporary connection is established and cleared in a manner similar to the control of a CS connection. This allows users to communicate by user–user signalling without setting up a CS connection, by utilizing the USER–INFORMATION messages once a temporary signalling connection is established.

As can be seen, there are a few alternatives to using the user–user signalling facility to identify the superchannels. However, the UUS facility of ISDN is a supplementary service, and as such may not be provided by all ISDNs. In this case, a different solution may need to be found. (See Chapter 7.)

6.5.2 A protocol over user–user signalling for superchannel identification

Superchannel identification between end-users can be achieved, without the co-operation of the network, by defining a new protocol whose message and information elements can be carried across the network transparently using the user–user signalling facility of CCITT Rec. Q.931. The new protocol defined below will uniquely identify a superchannel operating between the two pairs of communicating users. The general format of the protocol message is shown in Fig. 6.5. The meanings of the message fields are given in the key accompanying the figure.

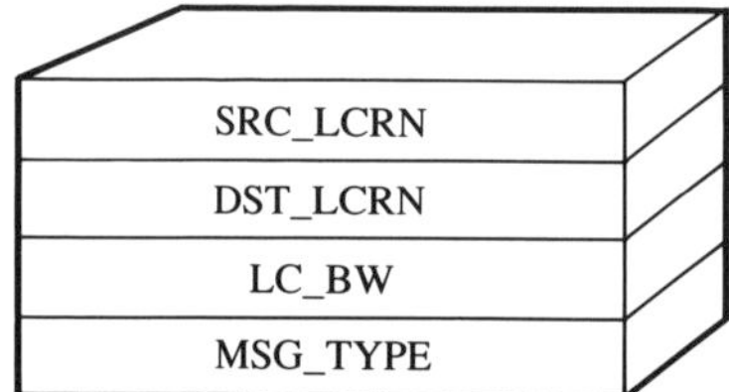

SRC_LCRN: source logical channel reference number; ISDN_ADD + reference no.
DST_LCRN: destination LCRN (not used until other end assigns and returns a value)
LC_BW: logical channel BANDWIDTH
MSG_TYPE: message type (LC_SETUP or LC_INC (logical channel SETUP or INCrement).

Figure 6.5 Superchannel identification protocol message format

It is assumed that the user–user signalling during the call SETUP is invoked for superchannel assignment and identification (service 1). If user U_1 wants to set up a B-channel connection to user U_2, with an option of extending this to a superchannel of size x BBUs, it must indicate this within a U–U message which will include a superchannel identification protocol data unit (SI_PDU) composed of a message as shown in Fig. 6.5 with MSG_TYPE set to LC_SETUP. This means that a superchannel association can be set up during a usual B-channel connection set-up by using the new protocol information carried transparently through the network in the U–U signalling message.

Channel identification Each co-operating user must be able uniquely to identify a superchannel at its UNI. Furthermore, this identification must be unambiguous globally between all of the communicating users. For this, the composition of each logical channel reference number (LCRN) as ISDN number plus a locally assigned reference number (distinct from the call reference value assigned for the physical channel) is suggested.

Operation The first message of this protocol will carry type LC_SETUP in its message-type field. Upon receipt, U_2 will record SRC_LCRN (the DST_LCRN field will be empty at the LC_SETUP request time) and create a local LCRN for future reference. A mapping of the two LCRNs will be kept by each user. If U_2 cannot provide the number of BBUs indicated by the LC_BW element within the SI_PDU, it will return the nearest amount available within its reply LC_BW. This will give an indication to U_1 about the congestion at the U_2 interface. This is registered as the 'bid value of bandwidth' at the destination and could be used in future for authorization of new call set-ups to satisfy the initial demand. Future increases in the logical channel bandwidth (i.e. by setting up a new connection) must be done with a SETUP message carrying an SI_PDU with its msg_type field set to 'INCrement'. In this case, both the source and destination LCRN values will be indicated within the SI_PDU. These will be used to associate the new connection with an existing superchannel. The LC_BW field will have no significance when INC message is sent.

When a physical channel is removed, no UUS is necessary to indicate the reduction of superchannel bandwidth, because each user will keep a mapping of the physical channel reference value to logical channel reference values within their respective interfaces (see Fig. 5.3).

Other possibilities The U–U signalling facility in ISDN provides a flexible and powerful method of user communication. It can also be used in channel reservation and action negotiation schemes. For example, if 2 H-channels exist between the two users, and a reduction of 4 BBUs are to be implemented, the two sides could negotiate the existence and reservation of 2 B-channels in advance of the removal of an H-channel. Furthermore, if the service 3 type of UUS is used, it is possible to associate two or more separate connections (in active state) with the same destination within a superchannel whenever such a requirement arises and to remove the superchannel association when the requirement has disappeared, carrying on the operation as separate channels, without the need to disconnect any existing connection (e.g. variable bandwidth voice/image communications).

6.5.3 Modification of CIIE

This method necessitates some changes to the format and usage of the current message/information elements. A new field needs to be created in the channel identification information element (CIIE) for the logical channel reference number. Also, the CIIE must be made an optional information element for the DISConnect and RELease messages (see Fig. 6.6). Again, a LCRN is assigned at the time of SETUP and is carried across to the other end in the CIIE.

6.5.4 Modification of CRIE

This method, again, necessitates changes to the CRIE. A new field to indicate a logical channel reference number must be added to the existing information element. This could conventionally be made to be the last byte of every call reference information element. Its usage is similar to the previous case (see Fig. 6.7).

8	7	6	5	4	3	2	1	Bit/octet
0	Channel identification 0	0	1	1	0	0	0 Information element identifier	1
Length of channel id. contents								2
1 ext	Int. id. pres.	Int. type	0/1 *	Pref./ excl.	D-ch. ind.	Info. ch. select.		3
0/1 ext	Interface identifier							3.1
1 ext	Coding standard		No./ map	Channel/map elt. type				3.2
Channel number/slot map								3.3
0/1 ext	Logical channel reference number (LCRN)							3.4

id.: identification; int. id. pres.: interface identification present; int. type: interface type
Pref/Excl: Preferred/Exclusive; D-ch ind: D-channel indicator
Info. ch. select: information channel selection; ext.: extension bit; elt.: element

NOTE*: Octet 3/bit 5 is spare. It could be used to indicate the existence of an LCRN byte. Alternatively, bit 8 of octet 3.4 could be used, leaving 7 bits for LCRN assignment.

Figure 6.6 Details of a new channel identification information element

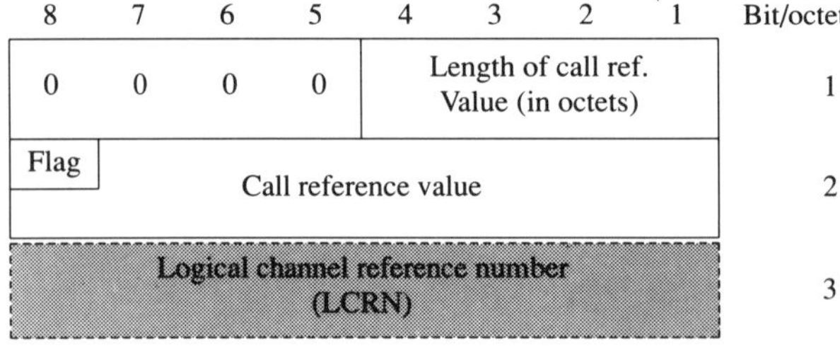

8	7	6	5	4	3	2	1	Bit/octet
0	0	0	0	Length of call ref. Value (in octets)				1
Flag	Call reference value							2
Logical channel reference number (LCRN)								3

Figure 6.7 Details of a new call reference information element

6.6 DYNAMIC BANDWIDTH VARIATION

The bandwidths of physical channels cannot be varied after their set-up except, it is assumed, in the case of superchannels formed by the *n* x 64 kbps method, where the logical and physical channels are identical. Several methods for bandwidth variation of a superchannel can be identified depending on the method used for its construction.

6.6.1 Aggregated superchannels

This method of superchannel formation implies the aggregation of several physical channels and necessitates the provision of TSSI within the customer premises (user equipment or CPE). End-to-end superchannel identification can be achieved with or without the network's active co-operation. If TSSI is provided by the CPE, some form of superchannel identification is also needed at the user side of the local interface. This could take the form of Q.931+ protocol described previously. However, in order to simplify dynamic bandwidth variation requests, a new Vary Bandwidth message can be used within the local interface (defined as Q.931++). This is identical to the one defined in the next section. In contrast, the network side of the local interface needs to identify the superchannel on an end-to-end basis. It does not need a Vary Bandwidth message, as the increase or decrease of the superchannel bandwidth will be achieved by the setting up of new or removal of existing connections on a one-at-a-time basis.

Any changes in the logical channel bandwidth must be done with reference to the underlying physical channel(s) in order not to upset the Q.931 call reference number assignment conventions. Current Q.931 protocol has no provisions for dynamically changing the bandwidth of a physical channel. This means that one may need to disconnect an existing physical channel in order to change the bandwidth of an associated logical channel.

When a new channel is to be added to the existing aggregated superchannel, its logical call reference numbers must be passed across the network to the other end in order to identify the new connection as belonging to the existing group. As mentioned before, this can be achieved by the use of UUS or the modified CIIE or CRIE, depending on whether the network is oblivious or informed about the association.

In the case of UUS, this applies to the SETUP only, since when an individual connection is to be removed it can be identified uniquely by the physical channel reference number (current call reference information element). However, optionally it could be repeated in the DISConnect message. The two ends can then adjust the bandwidth value of their local logical channel (superchannel).

The bandwidth variation is limited by the quanta of channel sizes available. For example, if a logical channel (LC) is composed of (1 x B) + (1 x H) channels, and it is required that bandwidth of size 2 x B is to be removed from the LC capacity, then one can either remove the 1 x B channel or the H-channel as a fixed, integral unit, since the UNI recognizes only these physical channel sizes. In the latter case, new connections of size 4 x B may be reconnected. The ISDN interface must be able to distinguish between different connections by their PCRNs and their associations with the LCRNs. This is depicted in Fig. 5.3.

6.6.2 $n \times 64$ kbps superchannels

In the case of a superchannel formed using the $n \times 64$ kbps facility, some means of dynamically changing the bandwidth of a physical channel needs to be incorporated into the D-channel protocol Q.931. Two proposals for incorporating this are:

1. *Set-up message on an existing call* This could be used with the previously assigned call reference number indicating a particular connection but with a new CIIE containing a modified time slot map. The TSs not already connected would then indicate their addition to the existing bandwidth. Alternatively, missing TS numbers would indicate their deletion from the existing assigned bandwidth.

2. *Vary Bandwidth message* A new message called Vary BandWidth (VBW) could be used to indicate that an increase or decrease is requested on an existing superchannel's bandwidth. The use of this message can be any time after a connection is established. This message could be either of the 'call information phase' or the 'miscellaneous' message type. Its suggested content is shown in Fig. 6.8. It will also necessitate a new information element called BANDWIDTH to indicate the size of bandwidth change. This is shown in Fig. 6.9. The indication for successful or failed attempt also needs to be defined.

Information element	Direction	Type	Length (octets)
Protocol discrimin.	both	M	1
Call reference	both	M	2–*
Message type	both	M	1
Channel ident.	both	O	2–*
Bandwidth	both	M	3–4

Message type: Vary Bandwidth; significance: global; direction: both
M: mandatory; O: optional information element

Figure 6.8 Suggested VARY BANDWIDTH message contents

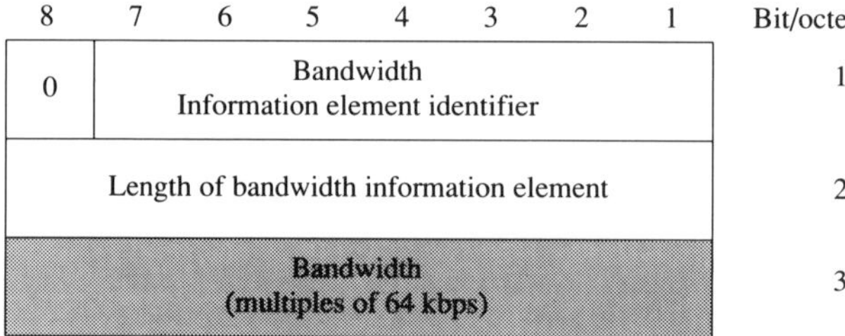

Figure 6.9 Details of new BANDWIDTH information element

6.7 CONCLUSIONS AND SUMMARY

In data applications over the ISDN, larger bandwidths—greater than the currently available fixed-size basic channel types—will be necessitated because of the packet and burst nature of data traffic. A further requirement at the user–PRISDN interface will be that of dynamic channel management, in order to manage the bandwidth allocated to different users dynamically and economically. Current CCITT protocols do not provide a separate facility for superchannel formation by physical channel aggregation or dynamic variation of bandwidth of an existing connection. Both of these shortcomings can be remedied by suitable modifications to the Q.931 protocol.

It has been shown that a superchannel can be formed by the aggregation of basic types of channels and treating them as a single *logical* channel. Here, it is assumed that TSSI within superchannels is provided by one of the viable proposals referred to in the literature. It is also shown that superchannel formation may be achieved on an end-to-end basis between the users, while the network does not need to know about this logical association of underlying basic channels. The case where the network assists in the superchannel identification may not be a practical solution, but the modified call reference and/or channel identification information elements may be used between user equipment and CPE for the same purpose, as a proprietary protocol. Furthermore, it is shown that the addition of a new message type to the D-channel protocol Q.931 could simplify dynamic bandwidth variation and lead to efficient dynamic channel management.

CHAPTER

SEVEN

SOLVING THE TSSI PROBLEM

In the previous chapter, we looked at the formation of superchannels ($n \times 64$ or nB + mH) and the protocols needed for channel identification and dynamic bandwidth modification. That work was based on the assumption that the time slot sequence integrity (TSSI) problem could be solved. The TSSI problem occurs because of the differential delays suffered by individual channels (or time slots) making up a superchannel as they travel through different switches and transmission trunks. In this chapter we deal with one of the approaches to the problem identified in Chapter 6 for time slot sequence integrity, i.e. the solution within customer premises.

The solutions of the TSSI problem can be classified according to several criteria. First, we can talk about the *type* of solution provided: this can be either TSSI avoidance or TSSI provision. Second, we can classify solutions according to the *layer* in which they are implemented in the hierarchy of the OSI reference model (ISO 1984). It should be noted that most of the solutions proposed in this chapter pertain to the use of independently switched or leased multiple (physical) circuits as a single wider-bandwidth transmission channel.

7.1 TSSI SOLUTION TYPES

Indications of how to avoid the TSSI problem have already been given in Sections 4.5, 4.7 and 5.1. Accordingly, and given the above comments, we can classify the TSSI problem solution in two broad categories:

1. *TSSI avoidance* Section 4.5 presented the resultant queuing system for different types of time slot utilization in a TDM frame; Section 4.7 mentioned the use of X.25 multi-link procedures (MLPs) over multiple physical circuits; and Section 5.1 discussed the effect of different queuing systems on the utilization of the wide-band channel. Put another way, the TSSI problem can be avoided in communications over multiple, distinct, switched or

non-switched physical circuits (e.g. multiple B-channels) by the use of facilities and features provided by the network and transport layer protocols. However, as pointed out earlier, this means that the aggregated physical circuits (or time slots) cannot be used as a 'fat pipe'. These points are further discussed below.

2. *TSSI provision* This involves the use of some algorithmic method, special training sequence (signalling) or framing technique so that the sequence of the data stream transmitted is kept in order or re-established at the receiving end. It is usually implemented at the data link or physical layers. The advantage of this method is the use of the aggregated wider-bandwidth channel as a 'fat pipe'.

7.2 LAYERING THE TSSI SOLUTIONS

The various solutions to the TSSI problem can be categorized according to the OSI layer to which they correspond. Currently, solution proposals are available in the transport, data link and physical layers. Each solution has its own pros and cons. A comparison is given at the end of the chapter.

7.2.1 Transport layer solution

This is a TSSI avoidance method and relies on the fact that one of the functions assigned to the transport layer is the re-sequencing of data packets (ISO 1986). In the case of a transport protocol running on top of a datagram network (e.g. TCP running over IP in the ARPANET), this feature is invaluable and necessary, as the packets travelling through such a network can inadvertently be received out of sequence. This is illustrated in Fig. 7.1. Packets sent over different B-channels may be delayed differentially, in which case the packet re-sequencing function of the transport protocol is activated. The other relevant feature of a transport protocol is the ability to multiplex/de-multiplex to and from multiple network connections (connection oriented mode). However, this does not apply to physically distinct resources (ISO 1986).

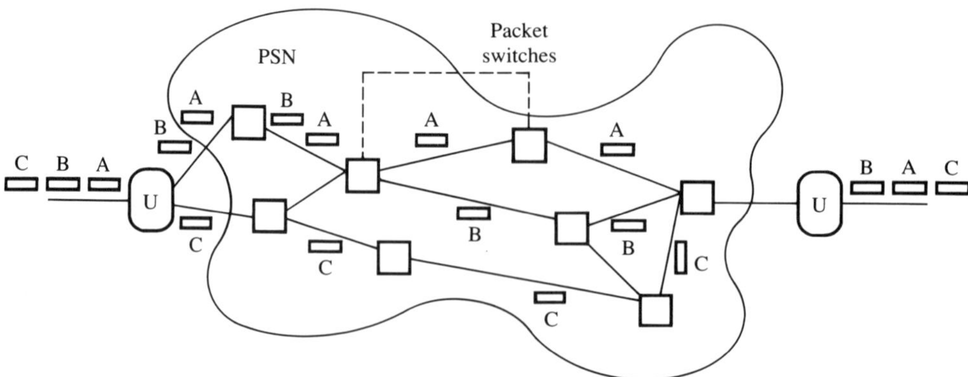

PSN: packet switching network
U: user equipment (e.g. gateway)

Figure 7.1 Packet reordering in a packet-switching network (connectionless mode)

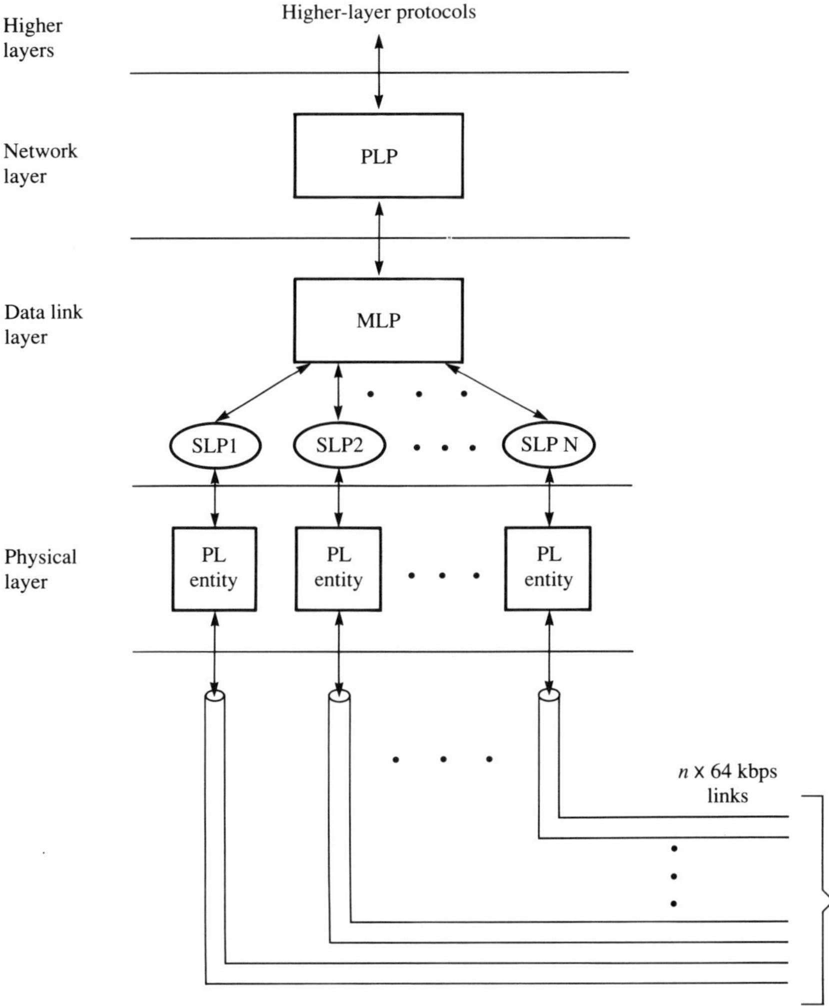

Figure 7.2 The use of multi-link procedure with N single-link procedures over $n \times 64$ kbps links

7.2.2 Data link layer solutions

The solutions available at the data link layer are the multi-link procedure (MLP), algorithmic scheduling and inband marker solutions.

Multi-link procedure (MLP) solution This is a TSSI avoidance method and involves the use of X.25 multi-link procedure (MLP) over multiple single-link procedures (SLPs) (CCITT 1988a). This is a subscription-time selectable option of the X.25 layer 3 functionality where the SLPs are different HDLC (or LAP-B) procedures running on each circuit. The MLP exists as an added upper sublayer of the data link layer. The MLP manages the traffic across n SLPs by adding control information to the X.25 packet header for packet numbering, sequence control and MLP reset. A multi-link procedure performs functions of accepting packets from the network layer (packet level protocol—PLP), distributing these packets to the available single link procedures and re-sequencing packets received from the SLPs to the network layer. Figure 7.2 shows the operation of the MLP over n SLPs. The MLP controls the transmission by the use of a window of width W. A similar window is used as a receive window. Assuming that the latest acknowledged frame is j, then packets numbered $j+1$ to $j+W-1$ can be transmitted. Acknowledgments received by the SLPs are passed to the MLP. Figure 7.3 shows the structure of the multi-link frame format.

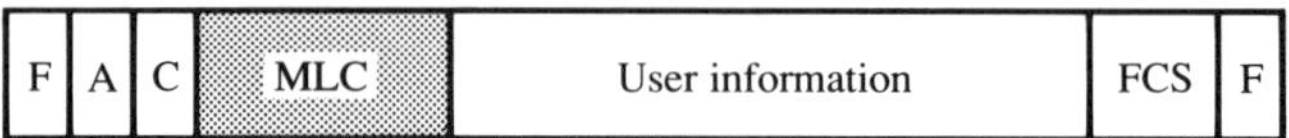

F: flag sequence; A: address field; C: control field
MLC: multi-link control; FCS: frame check sequence

Figure 7.3 Multi-link frame format

The packets can be dispatched to the SLPs in various ways; for example, round robin (cyclic) or load-adaptive algorithms may be used. It has been shown by Boltz *et al.* (1989) that cyclic distribution is less efficient than adaptive distribution. This is due to *window exhaustion* if any one link fails, and packet starvation can be experienced by errorless links arising from the error recovery process on the link with error. The adaptive packet distribution surmounts this problem since links in the state of error recovery do not receive new packets for transmission. Cyclic distribution is acceptable when link error probabilities are low. The complexity introduced by the adaptive algorithm may be justifiable only at high link failure rates.

Algorithmic scheduling solution This is a TSSI avoidance method that can be implemented at the interface of the data link and physical layers. It involves the use of special sending and receiving modules at each end. This is depicted in Fig. 7.4. The *load adaptive distribution* method (Altarah and Motard 1989) can be used as a frame scheduling technique for dispatching the data link frames to the appropriate channels of an aggregate $n \times$ B-channel. The data frames are recognized by their synchronization pattern (e.g. the HDLC flag) and can have variable lengths. This method relies on the use of sending and receiving counters, one per channel and direction, which keep a count of the number of bytes transmitted or received. The scheduling algorithm works on the principle that the next send channel is the one that has transmitted the least amount of data (in bits or bytes). The receiving module reassembles the data units received from the n channels using the same principle so that the data sequence is maintained. The next channel from which the data is used to reassemble is the one with the least amount of data already reordered.

It has been proved that the load adaptive distribution method performs better than a cyclic distribution. It has been shown to have two main advantages: data units need not be numbered, so that associated overheads are avoided; and it provides good performance under bursty traffic conditions (Altarah and Motard 1989).

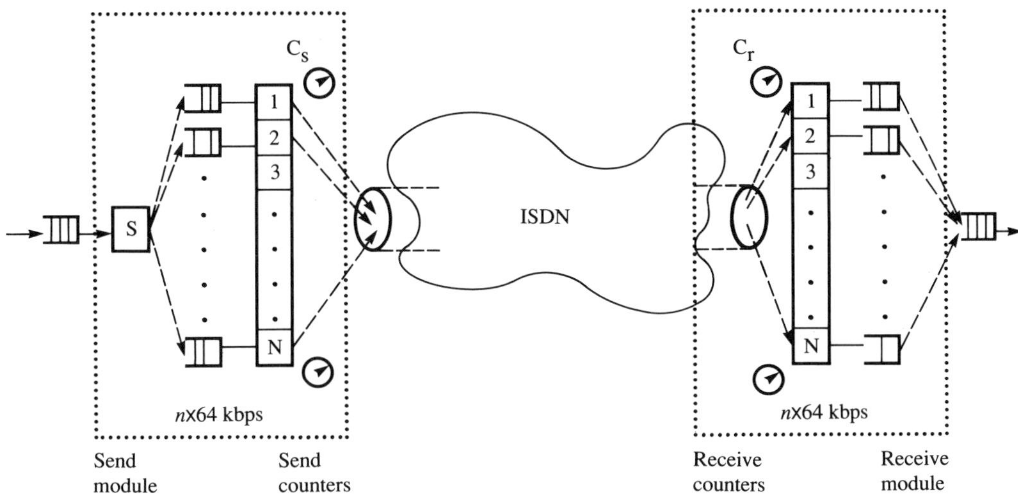

S: (frame) scheduler

Figure 7.4 The load adaptive distribution solution to the TSSI problem

Inband marker solution This is a TSSI provision method that is based on the transmission of special marker signals in a predefined sequence and in both directions in an in-band fashion over the B-channels to be aggregated. These marker signals are recognized and their relative positions in the data stream are noted at the receiving end. The receiver then calculates the displacements that will restore the signals to their correct ordering. This information is used in rearranging all the received data on these channels to produce a perfectly sequenced data stream (Burren 1989). The method is suitable for fixed $n \times 64$ kbps or $n \times$ B-channel aggregation as the marker signals are transmitted only at the beginning of a wider-band channel set-up. It can, however, be used to create a wide-band channel whose bandwidth is dynamically variable. This can be achieved by sending the marker signals periodically (i.e. once every rth frame) (Burren 1989). Markers use three single-bit flags for indicating slot synchronization achieved, slot ready and slot deleted. Slot synchronization bit makes it possible to indicate that the receiver has located markers in slot; slot ready indicates the successful incorporation of a time slot in the wide-band channel; and the slot delete indicates the removal of a time slot from the aggregate channel. The method can deal with both the slot rearrangement within frames and the skew between slots (see Fig. 7.5).

According to the method, a marker signal indicating the slot number in which it is transmitted is sent over each of the time slots that are to be aggregated. The receiving station can determine the skew associated with each slot by noting the relative position of the frame in which it is received. Similarly, the slot rearrangement within frames is determined from the slot numbers carried by the markers. A received slot versus frame matrix is kept at each end where the jth column is holding octets from the jth frame and each row holds the octets received in that slot. The orientation of the octet matrix is not of particular importance. What is important is the way in which the octets are assembled. The rearrangements are sorted out by scanning the rows (slots) in the order of transmitting slots (i.e. ascending order) as

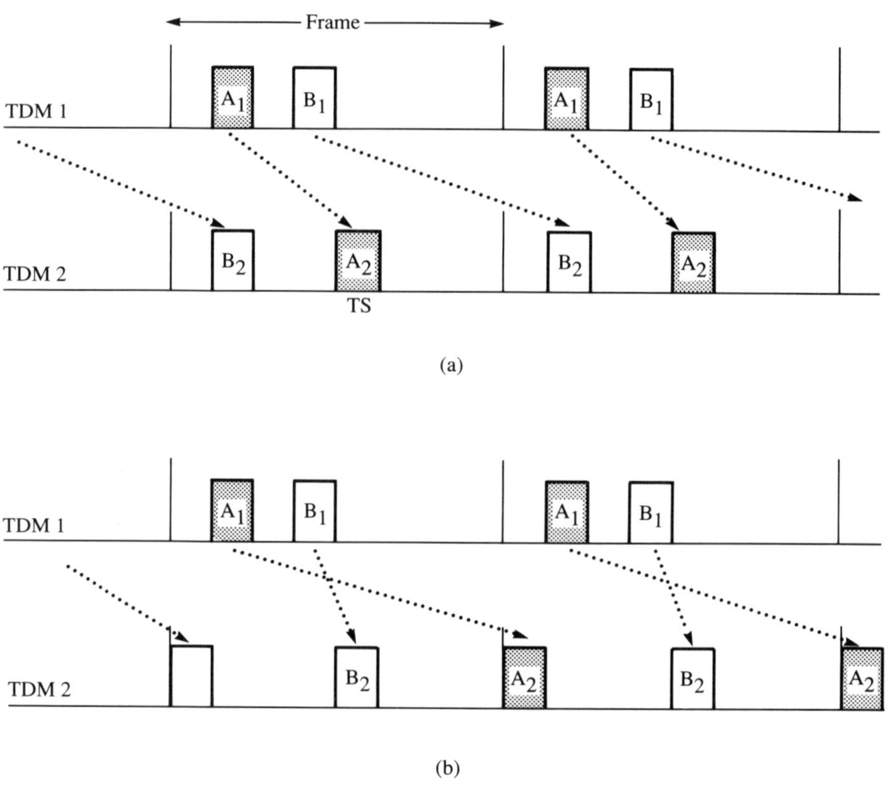

Figure 7.5 (a) Time slot skew (preserved sequence) and (b) time slots rearrangement (inversed sequence) in TDM networks

determined by the marker signals. Skew can be dealt with by taking octets from different columns (frames) of the octet matrix such that they represent data from the same scan. On arrival, a new frame is added to the matrix from the left-hand side and the time slots are taken from the right-hand side. Figure 7.6 shows the frame/octet scanning procedure.

Practical implementation of this method using Inmos transputer technology has been reported in the literature (Burren 1989). An overhead of 1 per cent is reported as necessary for marker framing and transmission. A rate of 12 000 packets/s (72 bytes/packet) can be handled in duplex form by the designed equipment. Slot synchronization is reported to take less than 100 ms, with a worst-case time of 200 ms. However, not all 30 B-channels in a primary-rate interface can be synchronized at the same time.

7.2.3 Physical layer solutions

These are collected in two groups. The first involves the use of some hardware techniques and the second relies solely on framing techniques. The latter method is really to be implemented at the boundary of the physical layer and the data link layer. Although I have placed it under the physical layer, some may argue that it is a sublayer of the data link layer. There are four known proposals for the framing method: CCITT Rec. H.221 (CCITT 1990); multi-frame

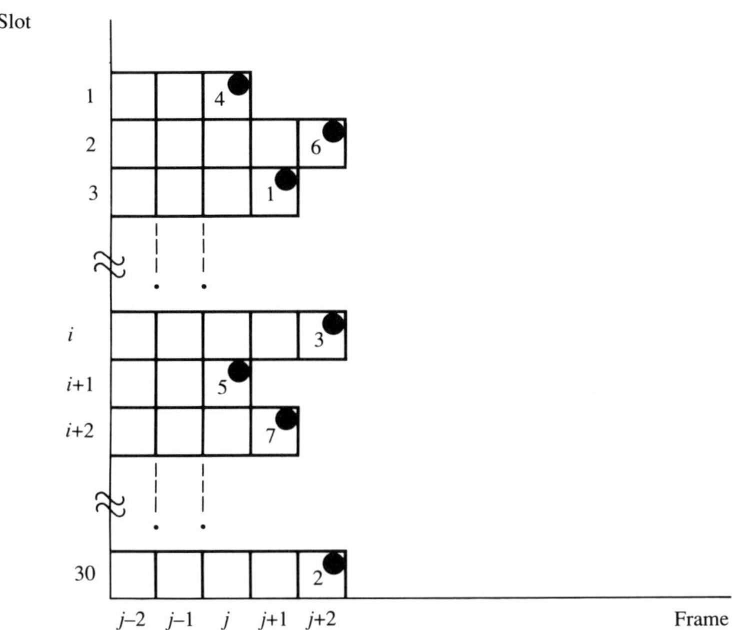

Figure 7.6 Frame and slot scanning procedure for inband marker solution

synchronization solution by Boltz *et al.* (1989); the draft Australian Standard DR 91140 (Australia 1991); and a recent proposal by the Bonding group in the United States (Bonding Consortium 1992). These are presented below.

Combined hardware and software solution This can be achieved by sensing the differential delay in the time slots belonging to an aggregate group using a training sequence and introducing compensating delays via buffering. By this method, both the frame delays and slot rearrangements can be detected. The software representing a finite state machine is placed into a programmable gate array in micro-code. Time slots from the faster channels can be delayed in a first in–first out (FIFO) buffer until the time slot from the slowest channel arrives. Similarly, time slots can be swapped according to the detected sequence by the state machine.

CCITT recommendation H.221 This is a well established international standard originally proposed to provide a framing structure for audiovisual teleservices in single or multiple B or H_0 channels or single H_{11} or H_{12} channels. The Recommendation provides a method for dynamically subdividing a transmission channel of 64–1920 kbps into constituent lower bit rate channels suitable for audio, video, data and telematics usage. The aggregate transmission rate is obtained by synchronizing and sequencing 1–6 B-channels or 1–5 H_0 channels or a single H_{11}/H_{12} channel. The principle is to use some of the capacity in the data stream for in-band signalling and negotiations. For a 64 kbps channel, an octet is transmitted at 8 kHz and each bit within the octet can be regarded as a sub-channel of 8 kbps. The eighth sub-channel thus formed is termed the *service channel (SC)* (CCITT 1990). The effective user data rate is therefore reduced to 56 kbps on each 64 kbps bearer channel.

The framing structure is composed of multi-frames, each of which is formed by 16 frames. A multi-frame is divided into eight sub-multi-frames each having two frames. Each frame is composed of 80 octets. Figure 7.7 shows the multi-frame structure used in Rec. H.221. A modulo 16 counter is used to number the multi-frames in descending order. The numbering is used by the receiving end in differential delay equalization and synchronization on separate connections. H.221 can deal with a differential delay of 1 s between bearer channels. The first channel established in an $n \times 64$ kbps aggregate channel is termed the *information (I) channel* and is used for controlling functions across the aggregate transmission. Some bits in the I-channel can be used for low-speed data transmission, giving an effective throughput of 62.4 kbps on each B-channel. A provision exists for 56–64 kbps interworking by 'rate adaption' so that the eighth bit is dropped in every octet. This creates a complex solution, since the service channel must now be placed on another sub-channel.

The service channel H.221 has a very rich repertoire of signalling facilities and incorporates error checking, signalling for control and maintenance. It also carries information enabling the selection of different applications such as data channels operating at 300 bps to 64 kbps (or up to 1920 kbps, if H-channels are used), multi-media communications by simultaneous video, audio and data at different rates, compressed speech, encoded video and still pictures, among many others. In addition, it has information fields for informing the other end about the standards used by the applications carried; for example, the CCITT Rec. H.261 is a standard for video encoding and is used over H.221 and H.222 channels.

A companion CCITT Rec. H.222 provides for channel aggregation of the H.221 type channels formed by $n \times 64$ kbps where n is 1–6. This effectively gives $n \times 384$ kbps channel aggregation.

Multi-frame synchronization solution This is a TSSI provision process where $n \times 64$ kbps channels that have been set up independently can be synchronized in such a way that variable delays on each link are taken care of at the receiving end. This synchronization is achieved by the use of a special frame structure. The transmitted data is divided into a multi-frame structure based on CCITT Rec. H.221 for multi-media communications. A multi-frame is created for each of the n channels to be used in aggregated fashion. This is shown in Fig. 7.7. Frames are transmitted simultaneously on all the circuits. At the receiving side, a synchronization unit is used to detect the frame boundaries, realign the frames and extract the transmitted bit stream. A multi-frame structure consists of 16 sub-frames each composed of 80 octets. Each sub-frame has two octets, dedicated to the frame alignment sequence (FAS) and bit-rate allocation signal (BAS), which identify the frame boundaries and assign the overall bit rate (used for multi-media applications). This constitutes a control overhead of 2.5 per cent. The user data is written across the frames of all links. In some implementations one bit of each octet is reserved for the synchronization overhead, leaving only seven bits for user data. This represents a 12.5 per cent overhead and needs $n + 1$ links to be set up in order to form a transparent link of capacity $n \times 64$ kbps. For cases where the switching and transmission system reconfigurations do not occur after the set-up of n circuits, the synchronization process can be done once at the start of the transmission, leaving all the bandwidth for data transmission (Boltz *et al.* 1989). On 64 kbps channels, each multi-frame has 160 ms of transmission time. Since on terrestrial links the maximum propagation time difference is 20 ms, the associated multi-frames on n channels are identified as those that

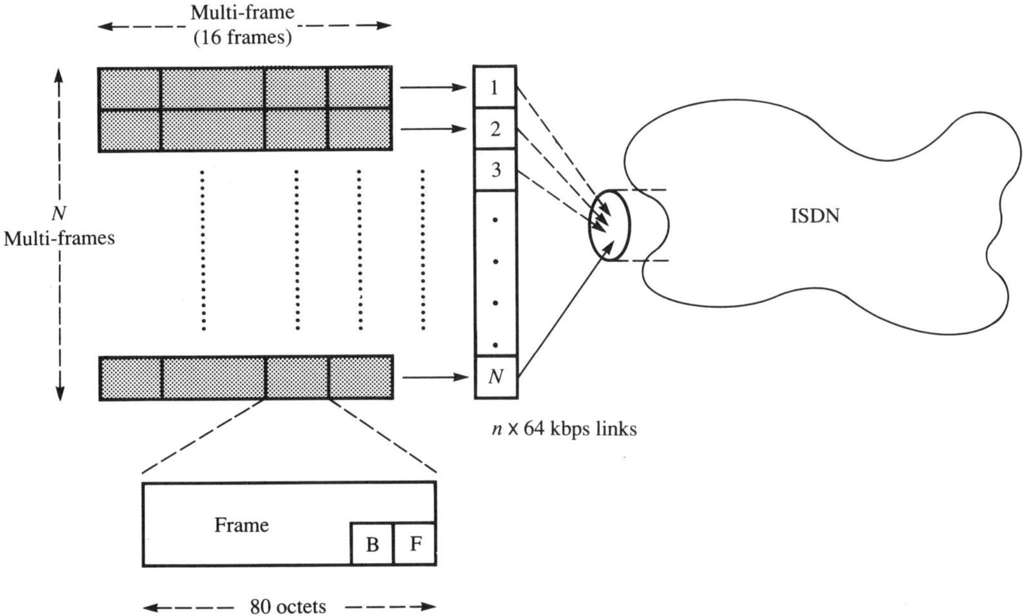

F: frame alignment sequence (FAS)
B: bit-rate allocation signal (BAS)

Figure 7.7 The multi-frame structure and transmission over n X 64 kbps aggregate channel

arrive within 20 ms of each other. The receiver then reassembles the transmitted bit stream by picking off transmitted octets from n multi-frames in the correct sequence. For links using satellite, numbering of the multi-frames is necessitated through the use of the FAS octet.

A proposed method of channel set-up is to establish a single B-channel which is to be used for channel negotiations between the two ends. Once this is settled, the remaining n–1 channels can be set up individually. In the third stage, synchronization frames are sent across each of the n channels in both directions. Once the frame alignment sequences are acknowledged, user data transmission within the multi-frame structure can begin. The BAS field can be used to indicate any addition or deletion needed to the number of channel used. Tests have shown that the observed quality of service (QoS) for the different data rates used was in accordance with CCITT Rec. G.821 (Boltz *et al.* 1989).

Framing solution II This is a TSSI provision method in which the data stream is divided into small segments and transmitted over multiple-channels. The n X 56/64 kbps working is achieved by defining a new service, a channel aggregation protocol, to be used in setting up the aggregated wider-band channel and a frame structure (Australia 1991). The frame structure is defined at the boundary of the data link and physical layers and provides a means of carrying the in-band framing and synchronization information, as well as spreading the data stream on a segment, octet and bit-by-bit base across the n channels. Further details of this method are given in Section 7.3.1.

Framing solution III This is again a TSSI provision method based on the superimposition of a framing structure over the $n \times 56/64$ kbps channels of an aggregate superchannel. The frame structure defined carries framing and synchronization information as well as superchannel identification information in an in-band fashion (Bonding Consortium 1992). Further details of this method are given in Section 7.3.2. A comparison of the three framing-based solutions is given in Section 7.4.

7.3 OTHER RELEVANT WORK ON TSSI PROVISION AND SUPERCHANNELS

Notable among other recent work on the TSSI provision and superchannel formation are the draft Australian Standard (Australia 1991) on digital channel aggregation, and the Bonding group's work in the United States (Bonding Consortium 1992). Both of these solutions have apparently been proposed to international standards institutions for adoption. At the time of writing, no information was available about their adoption.

The two proposed standards provide a solution to both the TSSI problem and the superchannel identification. Although the draft Australian Standard leaves the issue of dynamic variation of the aggregate channel (superchannel) bandwidth to further study, the Bonding Consortium proposal does provide a solution to this problem. However, neither of the two methods addresses the policies by which the aggregate channel bandwidth can be dynamically controlled. These are further discussed below.

7.3.1 Draft Australian Standard

A draft Australian Standard, 'Information Processing: Digital Channel Aggregation ($n \times 64$)' has been submitted to the ISO for adoption as an international standard. This standard addresses the problem of creating a high bandwidth digital duplex channel using multiple digital channels with lower bandwidths. Channel aggregation is achieved in the usual way by sending user data across the network over the individual lower bandwidth channels and at the remote end re-assembling and re-sequencing the data received (Australia 1991).

The standard deals with two users communicating in a point-to-point configuration using duplex aggregate digital channels. The standard does not address point-to-multi-point communications, nor the dynamic variation of the aggregate channel bandwidth by the addition and deletion of any number of B-channels during the course of the communications. The standard is applicable in aggregating bearer channels at both 64 and 56 kbps.

The number of B-channels to be aggregated is set up individually through the network. At the negotiation phase of the set-up, the compatibility negotiation may take place between the users over the D-channel using the lower layer compatibility (LLC) information elements. Alternatively, this may be done in-band using the first B-channel that is established.

Timing for the bearer channels is provided by the use of pseudo-random binary sequence (PRBS), allowing the equalization of differential channel delays and hence achieving the re-sequencing of the received bit stream. The polynomial used to generate the PRBS is $x^{10}+x^3+1$. This generates a ten-bit sequence, 0001110001 to 1111111111. The value of the frame bit in the *k*th sub-frame is the same for all channels and has the value of the *k*th term in the PRSB. The protocol defined for the channel aggregation has an optional in-band HDLC capability for more advanced negotiation procedures. However, this is left for further study.

The standard PRBS provides a differential delay detection of up to 2.046 s. However, this can be extended to a maximum of 7 s by the use of an extra bit in the I-stream (see section on 'Framing method').

Operation modes Three operation modes are envisaged:

- *Mode 1*: user data rate is an integral multiple of 64 kbps
- *Mode 2*: user data rate is an integral multiple of 56 kbps
- *Mode 3*: user data rate is equal to the aggregated channel bandwidth less the framing and information framing overheads

In modes 1 and 2, the actual channel rate is rounded to the next nearest integer. For example, for mode 1 over 64 and 56 kbps channels this is given by $n = r \times 64/63$ and $n = r \times 64/55$, where n is the raw channel rate and r is the user data rate, rounded to the next integers. For mode 2 these are given by $n = r \times 56/63$ and $n = r \times 56/55$ respectively. This provides a method for adding an extra channel for the framing and information overheads while preserving the user data rate. For mode 3, $n = r$ in either case. However, the actual user data rate for a 64 kbps bearer channel is reduced to 63 kbps owing to the usage of one bit in every eight octets for the overhead bits.

Framing method A frame structure is superimposed on the physical channels and provides space for the transmission of in-band signalling and synchronization information as well as user data. The frame structure is shown in Fig. 7.8. Each frame on a 64 kbps channel is composed of 1023 sub-frames. Each sub-frame contains 16 octets, giving 130 944 bits per frame and a transmission time of 2.046 s. Two bits per sub-frame are reserved for framing (F-bit) and information stream (I-bit)—for in-band end-to-end signalling—leaving 126 bits for user data in each sub-frame. This results in a minimum framing overhead of 1.5625 per cent of channel capacity.

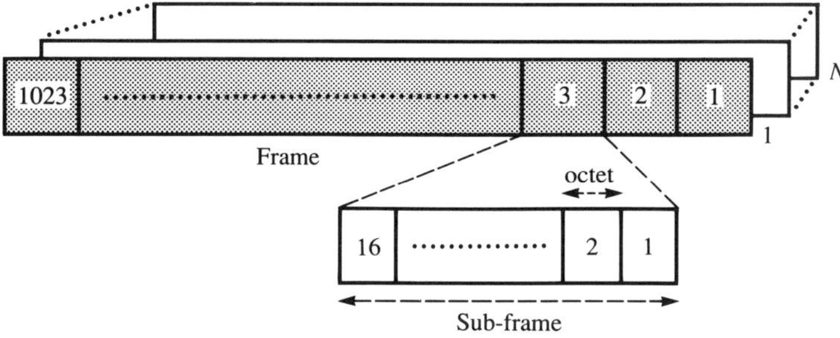

Figure 7.8 Framing proposed by the draft Australian Standard

The framing bit occurs in the first bit position of the corresponding octet of each channel sub-frame. For example, for the third channel in the aggregate, the framing bit appears in the first bit of the third octet. The information stream bit occurs on the $(8+i)$th octet of the ith channel sub-frame. The data stream is divided up into a sequence of fragments where the length of each fragment (measured in bits) is chosen to be one-eighth of the user data rate (measured in kbps). The fragments are then spread across the corresponding octets in each channel sub-frame. For example, a user data stream of 128 kbps to be transmitted over 3 x 63 kbps bearer channels (i.e. mode 1) will give a fragment size of 16 bits, causing a spread of some bits of the fragment onto the third channel since the F and I-bits occupy two bit locations in each of the channel sub-frames. This results in the fitting of data fragments i into octets numbered (i, n), where $1<n<3$ is the bearer channel number within the aggregate. In this case, spare bit capacity is left unused in the third channel. In mode 3, the data bits are spread across the (i, n) corresponding sub-frames of all the aggregated channels in a continuous and sequential manner, save for the F and I-bits encountered in each channel sub-frame. It can be seen that some wastage in bandwidth occurs in modes 1 and 2 as a result of framing and data bit assignment to sub-frames. This wastage can be represented as $1/(r+1) \times 100\%$, where r is the user data rate as a multiple of 64 or 56 kbps. In contrast, in mode 3 the only wastage is due to the framing and information bit streams; no spare bits are left unused, and the data octets are scanned across the channels in a sequential way and down the corresponding sub-frames.

Aggregate channel set-up An aggregate channel can be set up in two ways: using the D-channel signalling protocol with a modified lower layer compatibility (LLC) information element (LLC-IE), or using the in-band signalling method. In either case, an initial channel needs to be set up between the two ends in order to negotiate the aggregate channel establishment. This negotiation can be carried out either at the initial channel set-up or after its establishment in an in-band fashion.

1. *Outband signalling* This necessitates the use of a modified LLC-IE which is shown in Fig. 7.9. It contains fields indicating:
   (a) *Channel aggregation protocol* This is invoked by the use of this field in the LLC-IE.
   (b) *B-stream identifier (aggregate channel ID)* This is a unique identifier for the aggregate call and is used in subsequent negotiations and set-up or disconnect operations.
   (c) *Mode of operation indicator* This indicates selection of modes 1, 2 or 3.
   (d) *Channel identifier (within an aggregated group)* This is used to identify the bearer channels within the aggregate formation. The initial channel is always assigned the value of 1. The additional channels get numbers in sequence from 2 to n.
   (e) *P (number of bearer channels needed)* This is used in determining the value of n, the actual number of channels to be established for the aggregate channel. It is also used evaluating the availability of free channels.
2. *Inband signalling* This is used when attributes of an aggregate channel set-up cannot be negotiated using the LLC-IE. In this case, the negotiation is carried out through the use of the information stream F48 frames. The structure of this frame is shown in Fig. 7.10.

Additional channel set-up n-1 additional channel assigned to the initial aggregate call are established individually using the aggregate channel identification as a reference. In the case of out-band signalling, each SETUP message contains the LLC-IE as explained above. In the case of in-band signalling, the necessary information is carried in a frame structure carried over the information stream bits.

Aggregate channel disconnect This is achieved by the disconnection of the initial channel. All the other members of the aggregate channel are then individually disconnected using the ISDN call disconnect facility.

8	7	6	5	4	3	2	1	Octet no.	Description
0	LLC information element identifier							1	
Length of the LLC								2	
1	Coding standard		Information transfer capability					3	
0/1	0	0	1	0	0	0	0	4	64 kbps
0/1	0	0	1	0	0	0	0	4a	Optional
1	0	0	1	0	0	0	0	4b	
0	0	1	1	1	0	0	0	5	Channel aggregation protocol
0	S_0	S	S	S	S	S	S	5a.1	Aggregate channel ID and mode
0	S	S	S_9	M_0	M_1	M_2	R	5a.2	
0/1	CF	C	C	C	C	C	C	5b.1	Channel ID or no. of bearer channels
1	C	C	C	C	C	C	C	5b.2	

Figure 7.9 The lower-layer compatibility information element (LLC-IE)

Channel identification Superchannel and bearer channel identification is made using the identifier bits in the I-stream. Up to 1024 superchannels (aggregate streams) and up to 8192 bearer channels can be uniquely identified using the 10 and 13 bits of addressing, respectively.

Error detection This capability is not available in the present standard.

7.3.2 Proposals of the BONDING Consortium

The Bonding Consortium in the United States (Bandwidth on Demand Interoperability Group) is made up of a group of equipment manufacturers mainly in the United States but also has some following in Europe. The group has recently proposed a private version of a framing technique and superchannel identification to be used in meeting the interoperability requirements for $n \times 56/64$ kbps calls (Bonding Consortium 1992).

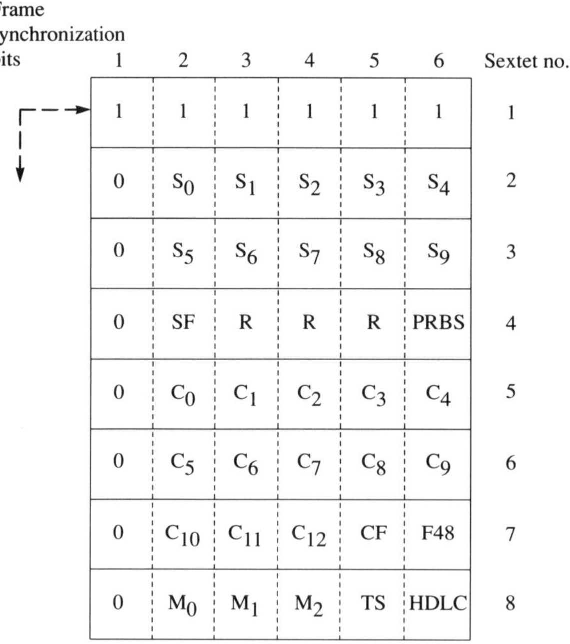

S_0–S_9: bit stream identifier bits; R: reserved bits
C_0–C_{12}: channel identifier bits; M_0–M_2: mode bits
SF, CF, F48, TS, HDLC, PRBS: individual status bits

Figure 7.10 The F48 frame structure

Main features

- The method is suitable for supporting applications with data rates of 8 kbps or multiples thereof, including $n \times 56$ and $n \times 64$ kbps, transmitted over 56/64 kbps bearer channels.
- The method provides automatic synchronization and alignment of multiple bearer channels.
- The method provides end-to-end negotiation for specifying the operating mode, bearer channel rate, phone number and other necessary information.
- Call and channel failures are detectable through monitoring functions provided. In the case of call failure, recovery procedures for call disconnect, rate reduction and bandwidth replacement (among others) are available.
- Framing overhead is 1.5625 per cent of the total bandwidth used.
- Dynamic varying of the transfer rate is possible during a data call; however, some data may be lost during the channel addition processes.
- Applications envisaged are: video conferencing, image transfer, LAN interconnection and bulk file transfer.
- The method is suitable for inverse multiplexing: a wideband connection can be split into multiple 56/64 kbps connections, transmitted and recombined at the destination to recover the wide-band information.

- Channel establishment and removal operations for the individual bearer channels of an aggregate channel are not covered by the standard: they use standard methods (e.g. ISDN call set-up).
- The physical layer electrical, mechanical and procedural (e.g. framing) details are left to the particular implementations.
- Higher-layer protocols using the service provided by this proposal are assumed to be implementation-specific.
- Up to 63 bearer channels of 56/64 kbps rate can be aggregated to form a superchannel of rate $n \times 56/64$ kbps.
- Overall transit delay for an end-to-end transmission is the sum of the longest channel delay and an implementation-specific fixed delay.

Frame structure Figure 7.11 shows the organization of the Bonding frame structure. Each frame is made up of 256 octets, of which four are overhead octets used in the establishment of the framing structure and for carrying negotiation information to the destination. A *multi-frame* is formed by the collection of 64 frames into a superstructure. The duration of a multi-frame is 2.048 s and the maximum delay equalization is possible for a relative delay of 1.024 s. Each bearer channel of the aggregate carries information using this framing structure.

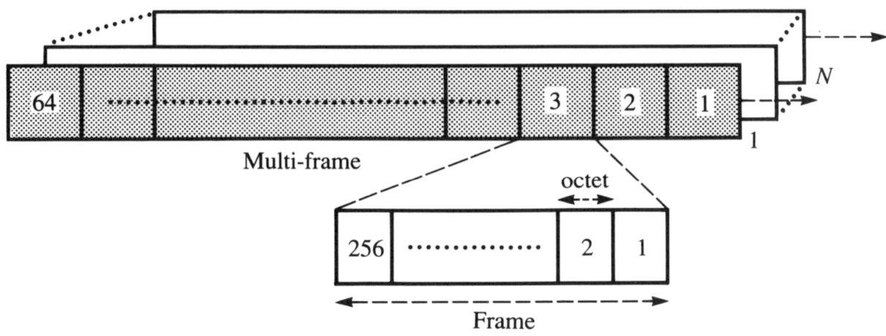

Figure 7.11 The Bonding Corporation frame structure

Within each frame, one of the overhead octets carries the *information channel* data. The information channel carries the aggregate-channel-related information including the channel and group identifications, telephone numbers for the channels and other negotiated parameters. User data is interleaved with the overhead octets in such a way that the user data spreads on an octet-by-octet base over the entire aggregate channel. The overhead octet in channel j is offset from the corresponding one in channel $j-1$ by 125 μs.

Operation The transmitting end sends user data in a special framing structure which is overlaid over the bearer channels partaking in the aggregate group. At the receiver side, all the channels are phase-aligned and synchronized using the framing structure. Data octets received on individual channels are restored to their original sequence and presented as a serial composite data stream at the output. This is done transparently to the applications carried as payload.

Channel establishment for the whole superchannel is achieved by first establishing a master channel, which is used to negotiate the aggregate call in an in-band fashion. The whole bandwidth is used for sending the information messages repeatedly. At the end of the negotiation, the call originator sets up the remaining channels belonging to the aggregate individually.

Four modes of operation are envisaged. Table 7.1 shows the comparison of different modes of operation.

Table 7.1 Comparison of Bonding mode functionalities

	Features					
Bonding mode	Initial parameter negotiation	Multi-frame structure on additional channels	Continuous delay equalization	Dynamic bandwidth change	Inband monitoring function[a]	Bandwidth available to users
0	√	x	x	x	x	Full available BW after negotiation
1	√	√	x	x	x	Full available BW after negotiation
2	√	√	√	√	√	$n \times 63/64$ of full BW (98.4375%)
3	√	√	√	√	√	Uses $n+1$ channels of which n are for user data

[a] Necessary for frame synchronization error detection.

- *Mode 0* Provides for the setting up of $n \times 56/64$ kbps call by initial parameter negotiation and then switches to the data transmission mode without delay equalization. No overheads are incurred after aggregate channel set-up and no dynamic channel bandwidth variation is allowed.
- *Mode 1* Provides for the setting up of $n \times 56/64$ kbps channels. It uses a training sequence at the frame synchronization phase to detect frame delays and time slot rearrangements, after which the inband signalling is removed, making full bandwidth available to the user. However, it cannot detect the loss of frame synchronization.
- *Mode 2* Supports the setting up of $n \times 56/64$ kbps channels, but, owing to the continuous use of overheads for in-band system monitoring, users can only get multiples of 63/64 of the bearer rate (i.e. 98.4375 per cent of the total bandwidth).

- *Mode 3* Supports the use of $n \times 56/64$ kbps channels (integral multiples of 8 kbps) where all channels support the same bearer data rate. The method uses $n+1$ bearer channels such that the extra channel capacity is used for the overhead transmission. The overhead octets are distributed over each of the bearer channels and provide in-band monitoring of frame synchronization as well as bit-error test using a cyclic redundancy check.

7.4 COMPARISON OF THE TSSI PROVISION METHODS IN CUSTOMER PREMISES

The transport layer solution does not necessitate any changes to the existing hardware capable of switching 64 kbps B-channels and the protocols used. However, owing to overheads involved in keeping a transport layer busy with correcting out-of-sequence packets on their arrival (and let's face it, in the worst case every one of them may arrive out of sequence), this method is thought to be very inefficient compared with other methods. Its main advantage is the ease of implementation. However, the behaviour of the transport protocol over a link whose bandwidth is dynamically changing needs to be studied further (Deniz and Knight 1989a).

The data link layer solutions are much more efficient than the higher-layer solutions:

1. The multi-link procedure solution is useful only in the case of the connection oriented mode of operation, and this is dependent on the use of X.25 suite of protocols. It still has a lot of overheads and will not be suitable in the case of LANs using the connectionless mode of data link layer protocols.
2. The algorithmic scheduling method does not need any framing overheads, but relies on the frame delimitation pattern. The suitability of its use in dynamic bandwidth management has been argued by Altarah and Seret (1991). However, the proposal has not been verified.
3. The marker method is adaptable to the dynamic bandwidth variation application since periodic marker frame transmission takes care of changes in the numbers of B-channels used. It relies on the fact that the transit delays associated with B-channels remain constant once they are established. The inband marker transmission inevitably takes away some of the available bandwidth. This unusable bandwidth can be as much as 1 per cent (Burren 1989).

Physical layer solutions are potentially the most efficient solutions to the TSSI problem:

1. For the combined hardware and software solution, few details are available at present as no proposals or products using this method have yet been announced.
2. The CCITT H.221 framing solution is a powerful method for achieving channel aggregation. It provides facilities for selecting a host of different applications to be run. It also has multi-media applications. It does, however, have some drawbacks. It has a high loss to overhead, and for this reason cannot provide a 64 kbps user rate. It also has a limitation to the number of channels that can be aggregated. The dynamic change of superchannel bandwidth is not possible, either. Table 7.2 shows a comparison of the different framing techniques that have so far been proposed.

Table 7.2 Comparison of different framing (aggregation) types

Feature	CCITT Rec. H.221	Australian draft Standard	Bonding Group proposal
Frame size (octets)	80	16	256
Multi-frame size (frames/octets)	16/1280	1023/16 368	64/3584
Capacity per channel (kbps)	56–62.4	63	63
Loss to overhead (%)	2.5–12.5	2.5–12.5	1.5625
Ability to provide 64 kbps user rate	No	yes—add extra channel or use HDLC	Yes—add extra channel or turn off monitor
Standard differential channel delay detection (s)	1	2.046	1
Extended delay detection	–	7	–
Error detection and correction	√	–	√
Octet synchronization	√	√	√
Maximum no. of B-channels	6	8191	63
Maximum no. of superchannels	–	1023	63
Parameter negotiation in B-channel	√	√	√
Parameter negotiation in D-channel	–	√ (using LLC-IE)	–
Dynamic bandwidth change	–	(under study)	√

3. The multi-frame synchronization solution is also adaptable to the dynamic bandwidth variation of the aggregate channel. However, it has a greater bandwidth wastage because of signalling and control overheads in comparison with the marker solution. As mentioned earlier, this unusable amount can be 2.5 or 12.5 per cent, depending on the implementation. Indeed, in one implementation this can be shown to be $1/(n+1)$, where n is the number of physical channels used in the aggregation.
4. The draft Australian Standard for TSSI provision (and channel aggregation) does not incorporate a dynamic bandwidth variation feature enabling any number of B-channels to be added or removed during the course of a communication session. Its other main disadvantage is that, in modes 1 and 2, $1/(r+1) \times 100\%$ is wasted because of framing and data bit assignment to the sub-frames used. The main advantages of this method are its suitability for 56 and 64 kbps communications and its applicability to both the ISDN and other types of multiple-link communications facilities.
5. The Bonding Corporation solution to the TSSI provision covers the multiple facets of the problem. It provides several levels of cohesion to the channel aggregation problem. Compared with the Australian draft Standard, it is much more detailed and covers the dynamic bandwidth variation (available in modes 2 and 3). However, it provides only an in-band signalling solution to the superchannel identification problem. This makes it usable in many more situations, but probably extends the initial negotiation time and hence channel synchronization.

7.5 SUMMARY

Time slot sequence integrity (or bit sequence integrity) is a problem when different channels are individually established, possibly through different switches and TDM trunks, and when these channels are to be used to form a wider-bandwidth channel. This problem can be solved in two ways: by circumventing it, and by time slot synchronization and sequencing. Problem circumvention is possible if different channels are treated as separate links and the packets or frames sent to the other end in parallel are reordered by the use of their sequence numbering at their destination. Channel synchronization and sequencing can be achieved by detecting the time slot differential delays and skew and placing appropriate delays on the data received on these channels at the destination. One increasingly popular and promising method is to use a framing structure to transmit segmented messages and reassemble them at their destination.

CHAPTER

EIGHT

QUEUE-BASED BANDWIDTH MANAGEMENT

As explained in Chapter 4, channel management (CM) is needed in order to control the setting up and removal of channels and to increase or decrease their bandwidth (bandwidth management) according to the traffic intensity and for some specified performance criteria. Chapter 5 described a dynamic channel management architecture which could implement the above requirements. Bandwidth management policies (BMPs) are needed in order to specify fully the action sequence under the dynamic behaviour of the system. In this chapter, four basic BMPs, based on *hysteresis threshold control*, using the queue length as the control metric and providing *multi-level service capacity control*, are proposed and their performances analysed using a simulation model (described in Appendices I and II) of the resulting queuing system. This is essentially a single superchannel, forming a variable service capacity M/M/1 queuing system to which the above-mentioned control doctrines are applied.

Section 8.1 describes the simulation environment and the experimental set-up. Section 8.2 presents the performance measures used, while Section 8.3 discusses the simulation models implemented; model assumptions, verification and validation are also covered. Section 8.4 presents the various bandwidth management doctrines that are considered in this book. Some heuristic multi-level BMPs are discussed in Section 8.5. The performance of BMPs is given in Section 8.6. Section 8.7 presents conclusions on the sensitivity and characterization of the results, and Section 8.8 summarizes the chapter.

8.1 THE SIMULATION ENVIRONMENT

8.1.1 Discrete event simulator

The Ethernet–ISDN relay (EIR) and ISDN simulator (ISIM) is an *event-driven* simulator implemented in the C programming language. The general 'event processor' used is based on the SLOANE (Satellite and Local Area Network Environment) simulation package (Sørensen

1988), while some 'queue management' routines are similar to the aforesaid package. The simulator can be used to generate packets according to different statistical distributions, define the network configuration, and gather statistics about the behaviour of the system under different test conditions. Using this test-bed, the performance of different channel management policies can be obtained. Figure 8.1 shows the general schematic of the simulator.

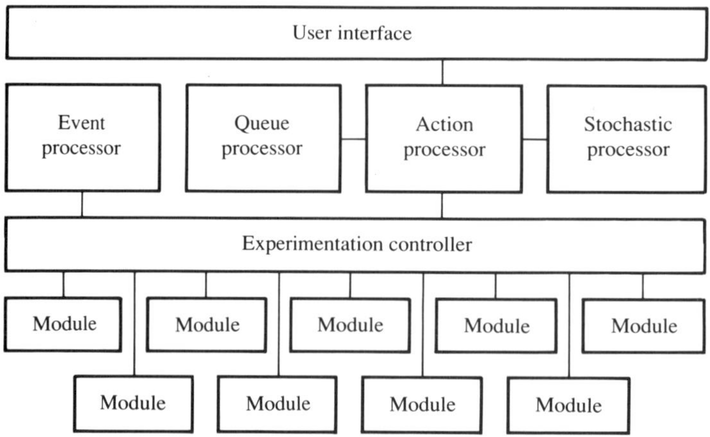

Figure 8.1 Elements of the Ethernet–ISDN (EIR) Simulator

The simulation model includes the following physical/logical modules:

- *User module* (UM—representing Ethernet and its hosts)
- *Relay module* (RLM—which is further divided to IPRP and IRP)
- *ISDN interface module* (IIM)
- *ISDN module.*

All except the last module correspond to a similar set of processes within the EIR architecture described in Chapter 5 (see Figs 5.9 and 5.10). A description of the simulation model is given in Appendix I, while Appendix II provides results used for the simulator validation. Timings for the D-channel signalling are obtained from a calculation of their message sizes and the signalling channel capacity of 64 kbps (see Appendix III).

8.1.2 The statistical nature of simulation outputs

A simulation process involves the stimulation of a *simulation system* (or mechanism, e.g. a discrete-event simulation of a real system) by a stream of random data consisting of a set of sequences of random variables whose distributions are predetermined. These random variables represent features of the work load or features of the system (e.g. packet arrival times and packet length distributions). The simulation mechanism acts on these input sequences to generate a random output that consists of a set of sequences of random variables such as throughput, delay or system response time. The output distributions are unknown. The main purpose of the simulation is to estimate certain characteristics of these output distributions.

In order to estimate the required performance parameters, the simulation experiment is run with different realizations of the input stream generated according to the prespecified distributions, and the corresponding output sequences are collected. Using these multiple realizations, a statistical inference is made to the required performance characteristics. Hence one can view the simulation as a set of statistical experiments, whose output must be statistically analysed in order to obtain the underlying performance characteristics.

8.1.3 Performance measurement

Two types of performance measurements are required in the analysis of simulation runs:

1. *Transient behaviour—short-run (finite-horizon) evaluation* Characteristics of the output of the simulation that depend upon the initial conditions and the point at which they occur, either in sequence or in time, are called transient characteristics. Here we are interested in the evolution of the system during a relatively short period of time.
2. *Steady-state (equilibrium) behaviour—long-run (infinite-horizon) evaluation* Here it is the system performance in the long run that is of interest. If a system has an equilibrium state, in the long run any transient effects should have settled down. Hence when the equilibrium state is reached, the probability of finding it in any given state is invariant with time.

The output analysis of a simulation is strongly dependent upon whether the system simulated is assumed to be a finite- or infinite-horizon system. Fairly conventional statistical methods apply to finite-horizon simulations (see Bratley *et al.* 1983). In this case the effect of initial conditions is reflected fully in the analysis and the serial correlation in a given run is irrelevant. For the steady-state simulations, however, the opposite holds. In that case, an estimate of infinite-horizon performance is obtained from a finite-horizon sample. This presents difficult problems, e.g. the serial correlation of samples and the existence and starting-point of the steady state, which are dealt with in the following sections.

It can be argued that most real-life situations lead to non-steady-state cases. However, most literature on simulation output analysis deals mainly with the steady-state condition. Again arguably, an exclusive focus on things like mean queue length or mean waiting time, steady state or otherwise, is naive. The ubiquity of this approach in the literature does not justify its automatic adoption (Bratley *et al.* 1983).

From the foregoing arguments, it can be seen that a judicious judgement of the applicability of either method is needed.

Transient performance characteristics Transient characteristics are of importance to an experimenter for two reasons. First, there are situations in which the transient characteristics of a system are the most important aspects, or, indeed, where only the transient behaviour may exist for a simulation. Second, these characteristics may be important in controlling the effect on the estimation of some steady-state characteristics. Furthermore, in this case the only way to get a sample of independent observations for a performance measure is to run the simulation n

times, over the same period and starting with the same initial conditions, but using different random number sequences. This is referred to as the *independent replications* method.

Steady-state performance characteristics Here, two major problems must be tackled. First, the effect of the transient phase must be eliminated since it introduces bias in the readings. Second, because of the long runs needed, one may not be able to afford to replicate the measurements, and hence readings from the one long run are taken to form the replicated samples. This occurs in the *batch means* method described below.

In the first case, the effects of the initial conditions (also known as the *warmup* effect) can be eliminated by discarding the data collected during the transient part of the run (i.e., the first l initial sample values). However, this itself introduces two further questions concerning when the transient part stops and the steady state starts, and the existence of a steady state. The answer to the first is that the transient phase does not end at a particular point, but the behaviour of the sequence gradually converges to the steady-state behaviour. The latter question is rather controversial and more difficult to ascertain. The steady state of a stochastic model is defined in terms of the *probability distribution* of the system, and not in terms of a particular value of the state. It should be recognized that the equilibrium is a limiting condition that may be approached but never attained exactly. There is no point in extending simulation time beyond which the system is in equilibrium. Hence we choose some reasonable point beyond which we are willing to neglect the error that is made by the equilibrium assumption (Kobayashi 1978).

The proper statistical analysis of steady-state simulations is still an unresolved problem (Bratley *et al.* 1983). However, unlike the case of transient characteristics, the estimation of steady-state characteristics can be made using several different approaches for sample data collection and analysis. These include the independent replications, batched means, regenerative, and spectral analysis methods. Each method has its advantages and disadvantages. The 'independent replications' in this case are similar to those in the transient case except that the treatment of the variance is slightly different (Lavenberg 1983).

The batched means method uses a single long run where, after deleting the warmup effects,[1] the run is divided into n portions or sub-runs and the samples of these portions are treated as multiple replications from which a statistical inference is made. The statistical problem faced is that a serial correlation between these samples exists and cannot be ignored. This means that the samples are not truly independent. A way round this problem is to make each sub-run so large that any dependencies are negligible. Another method is to take these dependencies into account, leading to a time-series analysis. A third way is to define sub-runs in such a way that the observations become independent; this leads to the regenerative method.

8.1.4 Experimental set-up

In our experimentation the independent replications method is chosen for simulation measurements. As already mentioned, this method is suitable for both the short- and long-run experiments. Three experimental details need to be clarified: the elimination of the warmup effect, the sample size to be collected (or run length), and the number of replications.

[1] In the batched means method, the warmup effects have to be dealt with only once, rather than k times as in the case of replication. Hence the batch means method is more efficient; i.e., fewer sample values may be needed to achieve a given accuracy.

Warmup effect As mentioned earlier, one major practical problem in the steady-state analysis is the determination of the extent of the transient phase for the warmup effect. Various heuristic approaches exist for this. One method is to make short independent replications of the simulation. From these M replications, the mean of the distribution is found and a test for its convergence is made. These can then be plotted together with a set of 90 per cent confidence intervals. This graph will show the trends in the mean value from which a decision on the amount of deletion can be made. Another approach, quoted by Lavenberg (1983), is the use of moving averages on M replications to smooth out the transients in the trend of the mean. Mitrani (1982) suggests that one could maintain a current estimate of the probability of a particular system state and examine it at selected points of time, assuming equilibrium when the difference between consecutive estimates becomes small. Alternatively, and more crudely, the transient phase can be said to have ended when the current estimate of some performance measure stops moving in one direction and starts oscillating. MacDougall (1987) suggests a rule-of-thumb deletion of 5–10 per cent.

In the experiments described here, which fall into the steady-state analysis category, a 10 per cent deletion is set up as a rule of thumb, although the linear correlation of a measured performance index (mainly response time in our experiments) with time is also performed, giving an added security against the transient phase effects in the long-run statistics. The Pearson product-moment coefficient of correlation (Mood *et al.* 1974; Tashman and Lamborn 1979) is defined by equation (8.1) and (8.2)). A value of correlation coefficient r $(-1 \leq r \leq 1)$, at or around 0 indicates little or no correlation between the performance index and time. In other words, the value-of-performance index thus obtained is assumed to be independent of time; i.e., it has approached its steady-state value.

$$r = \frac{\mathrm{Cov}\,(X, Y)}{\sigma_X \, \sigma_Y} = \frac{\Sigma(X-\bar{X})\,(Y-\bar{Y})}{\sqrt{\Sigma(X-\bar{X})^2 \;\; \Sigma(Y-\bar{Y})^2}} \tag{8.1}$$

where Cov(X, Y) is the covariance of random variable X and Y and σ_X and σ_Y are the standard deviations of the respective variables. For 'on-the-fly' calculation of r during simulation, the version shown in equation (8.2) is used:

$$r = \frac{n\Sigma XY - \Sigma X \Sigma Y}{\sqrt{n\Sigma X^2 - (\Sigma X)^2}\;\sqrt{n\Sigma Y^2 - (\Sigma Y)^2}} \tag{8.2}$$

where n is the number of samples.

Sample size It was found that a sample size of 100 000 packet transmissions provided a good compromise between accuracy and run time. Increasing this to 400 000 did not provide much more improvement either in the accuracy or in the serial correlation index. Hence, for the work described in this chapter, the experiments were run until 100 000 packets were transmitted by the source and received at the destination.

Number of replications Once the length (or duration) of each simulation run (replication) is determined by any one of the above methods, the number of replications to be used can be determined depending on whether a fixed run or a fixed confidence interval is desired. The following two basic approaches can be used in the collection of results:

1. *Fixed number of runs* The experiment is run until k replications are collected. From these replications, an overall mean and its corresponding confidence interval (CI) and accuracy (relative half-width) are calculated. This means that the CIs obtained for different points on a curve may have different values and accuracies. (See Appendix II.)
2. *Fixed confidence interval runs* This method necessitates the adjustment of the number of runs (replications) to be conducted according to the desired CI level. At the end of a minimum number of runs, the mean and CI of the values obtained is found. If it is below the desired level, then further runs are executed, each time calculating the new mean and CI. This process is carried on until either the desired CI value or the maximum number of runs allowed (usually set to avoid experiments running for an infinite amount of time) is reached.

The first method is simpler to implement and run. The second method, although providing more uniform CI values for each point on a graph, is more involved. The experiments described here have been conducted according to the 'fixed number of replications' method where the value of $k = 10$ for the number of replications has been used (except where indicated otherwise).

8.1.5 Analysis of results

For each group of k replications, an overall mean value and its confidence interval is found. From the half-width and the mean value, the accuracy (relative half-width) of the measurement is calculated. The main performance measures obtained are:

1. Mean number of customers in the system, N
2. Mean number of customers in the queue, Q
3. Mean response time (or sojourn time), RT

Additionally, the following values are collected over all the k replications:

- Mean number of switchings (per packet), SW
- Mean bandwidth used (in BBUs—64 kbps), BW
- Mean channel holding time at the BW value (ms/packet), T_H
- Mean bandwidth cost (BBU ms/packet), C_{BW} ($= BW \times T_H$)
- Distribution of bandwidth (BW) versus the number of switchings or the mean duration of switched time

For measures 1–3, confidence intervals are calculated; for the last five measures, only the mean values are calculated.

8.2 PERFORMANCE MEASURES

A channel management policy will operate to switch the required number of channels in and out as the load on the system varies. It will do these actions depending on how it is set to minimize or maximize certain performance criteria. In order to compare different control models and policies, measures of effectiveness need to be defined. One method is to superimpose a cost function on the performance of different models in order to account for the costs of packet servicing (including channel holding time and switching costs) and packet sojourn time, as is done in Gebhard (1967) and King and Shacham (1986, 1989) (see also Section 1.6.3 above). In fact, Gebhard assumes a cost of service based on the costs of bandwidth holding and switching, and treats the cost of queuing separately. King (1990) takes into account only the channel-switching costs and the sojourn time costs. However, one shortcoming of these methods is the fact that relative costs of customer servicing and channel-switching versus waiting (or sojourn time) are not in the same units; that is, while the cost of servicing and switching may be measured in money, the cost of delay is not so clear-cut. Additionally, in ISDNs the tariffing structure does not have switching costs (see Section 2.10). There are no guidelines as to how the variable bandwidth channels discussed in Chapter 6 are going to be charged by service providers. Assuming that the $n \times$ B channel formation will be available only and that each additional B-channel is going to be timed and charged separately, it will be costly if a policy achieves average B-channel hold time coincident with less than the median of a charge period. This loss will be proportionally higher if the total 'channel hold' is a few charge periods. A delayed channel close policy, discussed in Chapter 9, compensates for this charging problem. However, in this chapter we will not consider that problem.

It can be expected that some policies will perform better than others under different arrival and service processes. In general, the 'goodness' of any CM policy depends on the following factors: the mean and variance (spread) of bandwidth used, the mean bandwidth cost, the number of switchings necessitated, and the packet performance (mean delay D and throughput T). Hence, cost function C is given by;

$$C = f(BW, \sigma_{BW}, C_{BW}, SW, D, T). \qquad (8.3)$$

This cost function is a rather complex one to evaluate. Instead, one can use a simple average cost function which takes into account only the mean bandwidth cost C_{BW} and the response time RT of the form:

$$C_I = c_1\, RT + c_2\, C_{BW} \qquad (8.4)$$

and the average cost function, taking into account the number of switchings, would be:

$$C_{II} = c_1\, RT + c_2\, C_{BW} + c_3\, SW \qquad (8.5)$$

where c_1, c_2 and c_3 are the respective cost indices.

When multi-level service capacity control doctrine is used at an interface supporting (more than one) competing channels, other shortcomings are the absence of measures of effectiveness as to the fairness, responsiveness and burstiness of the policies used. Fairness gives a measure of 'greediness' (or rather the lack of it) for a policy. Responsiveness gives an indication of how fast a policy can respond to changes in the input (mean packet arrival rate). Burstiness deals with the ratio of the mean to standard deviation of the number of basic bandwidth units used. Admittedly, these measures are not easily obtainable and so are not pursued here.

Owing to the difficulties of assigning relative cost values to the various measures discussed above,[2] and in order to achieve some form of comparison between the different policies, in this book we use the following measures. The first measure of effectiveness considered is the bandwidth holding cost given by the *mean bandwidth cost*, C_{BW}, in units of

[2] Indeed, Gebhard (1967) stresses that 'cost allocations do not always make the best control option apparent'.

BBU milliseconds per packet. This is suitable, since policies with total channel closure (server shut-down) are considered here. Its value is obtained by using equation (8.6), where bw_i are the different bandwidth values held for a duration of T_h (bw_i) each (for the whole duration of the experiment); N is the maximum number of channels available, and k and m are the numbers of replications and total number of 'packet completions' in each replication, respectively. Note that, when a channel is totally shut down (i.e. when $bw_0 = 0$), then no channel holding time and therefore no bandwidth costs are incurred.

$$C_{BW} = \frac{\sum_{i=0}^{N} bw_i \; T_h(bw_i)}{k \quad m} \tag{8.6}$$

The second measure of effectiveness is the response (sojourn) time in units of milliseconds and is treated separately (in line with Gebhard's (1967) treatment of 'cost of queuing'). The third measure is the mean number of switchings per packet, *SW*, where the total number of switchings is calculated for the whole experiment and then scaled off. Finally, there is the variation of the mean number of customers in the system, N, with the mean arrival rate, λ.

8.3 SIMULATION MODELS

The basic simulation model (see Fig. I.3 in Appendix I) is the same for all four of the heuristic CM policies tested. The main difference occurs in their channel management decisions. Figure I.3 shows parallel queues, each with an assumed variable capacity server. In actual experimentation, only one such superchannel has been implemented at the ISDN interface. Consequently, the interesting but difficult case of the behaviour of several identical variable capacity queuing systems operating in parallel and sharing the total bandwidth has not been investigated. The basic resultant queuing system of the ISDN interface superchannel queue is of a G/M/1 type owing to the presence of a prior M/D/1 type queuing system in tandem with it (see Fig. I.4 in Appendix I). Since the (fixed) delay associated with the M/D/1 queuing system is almost negligible (10^{-9} s), the resultant model approximates very closely to the M/M/1 type performance. This has been proved by the validation experiments mentioned in Section 8.3.2.

8.3.1 Major model assumptions

The major model assumptions are given in Appendix I and include the main simulator timings used. Other issues are described below.

D-channel signalling delays The D-channel signalling transmission delays can be calculated by working out the size of basic messages used in the call set-up and removal phases for a circuit-switched connection. For a PRISDN interface served by a D-channel operating at 64 kbps, the sample calculations are given in Appendix III. Note that the processing load this presents for the interface processor is assumed to be negligible since it could be running on a dedicated signalling processor.

Channel switching times The B-channel switching times have been assumed to be exponentially distributed with parameter γ for the switch-in (on) times and instantaneous for the switch-out (off) times. The B-channel switching times include the signalling processor scheduling and processing delays, the D-channel signalling message transmission delays and the ISDN internal network switching delays. Owing to the D-channel signalling transmission delays for the SETUP message (2.5 ms), the minimum switch-in time has been assigned to be 3 ms, in order to avoid problems with B-channel switching (the random number generator could potentially return a zero value) when the arrival rate is high and the threshold value is low. Two options arise here: first, the assignment of a statutory 3 ms delay whenever the random number generator returns a switching delay of less than 3 ms; second, the re-selection of another switching delay from the random number pool until the switching delay returned is greater than or equal to 3 ms. The mathematical representations for the two cases are given below.

Case 1 Set $x = 3$ if $x < 3$; then

$$X = \begin{cases} 3 \text{ with probability } (1-e^{-3\gamma}) \\ x \text{ pdf. } \gamma e^{-\gamma x},\ x > 3 \end{cases} \tag{8.7}$$

Case 2 Re-select a new x if $x < 3$; then

$$P(X \mid x > 3) = \gamma e^{-\gamma(x-3)},\ x \geq 3 \tag{8.8}$$

In the rest of this book case 2 will be used.

8.3.2 Simulator verification and validation

Before any measurements can be taken using the simulator, the simulation model needs to be verified and validated. *Verification* refers to the comparison of the conceptual model to the computer code that implements that conception. *Validation* refers to the act of determining that a simulation model is an accurate representation of the real system (see Banks and Carson 1984).

Verification of the simulation model was carried out by building elaborate diagnostics facilities into the simulator code. These include the event/trace recording, module/action time recording, state recording, input/output recording and physical buffer recordings. The modular design of the simulator has also helped in the verification process.

Validation was carried out by testing the model against an M/M/1 system, Gebhard's (1967) model, and the model by Moder and Phillips (1962). The results of these tests are given in Appendix II. From these results, it was concluded that the simulation model behaved as it should under the test conditions. The final simulation model for the CM policies was also tested against an approximation technique (see Appendix I).

8.4 BM POLICIES—AN EXPERIMENTAL EVALUATION

The queuing models for channel management and threshold doctrines for bandwidth management have already been discussed in Chapters 1 and 4. In this chapter we look at the performance of several heuristic multi-level service capacity control doctrines based on *hysteresis threshold control*. We call these the bandwidth management policies, and they include total channel switch-off and assignment (start-off). The main threshold doctrines are listed as the point and hysteresis types. Gebhard (1967) showed that for the M/M/1 queuing system the point threshold performed worse than the hysteresis threshold control for all (service and queuing) cost combinations. Therefore, we shall limit our study to the hysteresis threshold control type of policy.

A further classification (the full taxonomy is given in Chapter 9) is based on the control levels: bi-level (see e.g. Gebhard 1967; Harita and Leslie 1989 for M/M/1 queuing system) and multi-level (see e.g. Moder and Phillips 1962 for an M/M/c queuing system) threshold controls. Since the ISDN interface composed of *n* x B-channels (or, indeed, *n* x 64 kbps) presents a facility that can be utilized in steps of 64 kbps, then a multi-level control policy is naturally suitable for this interface. In the multi-level control policy, the bandwidth of a superchannel is increased or decreased in steps of 64 kbps (i.e. a BBU). The maximum bandwidth allowed is specified at the beginning of the session (e.g. by a bandwidth allocation policy).

Since this study is through simulation rather than analytical techniques, I have built in various practical conditions. One of these is the channel switch-in delays discussed in the previous sections. The other is the actual point in time when any change in bandwidth is effectively implemented. This is affected at the *end* of the current packet transmission; i.e., if a packet is already in the server and is being transmitted, then the change in bandwidth becomes effective at the end of the current packet service. This, I believe, is a more realistic system behaviour than one where the bandwidth can be changed arbitrarily at any instant. Four heuristic policies based on the multi-level service capacity control and using the hysteresis threshold doctrine are described in the next section.

8.5 HEURISTIC BANDWIDTH MANAGEMENT POLICIES

The M/M/1 queue has a monotone-hysteretic optimal service capacity (rate)[3] control policy when switching costs exist, and a monotone optimal control policy in the absence of switching costs (Lu and Serfozo 1984). The hysteresis control form, where the switching costs exist, arises from the need to minimize the number of switchings and hence make it economical once an increase in service rate is achieved. The hysteresis service rate control where the switching occurs instantaneously is studied by Gebhard (1967). The case where the switching takes a time that is exponentially distributed has been studied by Harita and Leslie (1989). However, both of these studies consider only a bi-level change in the service capacity (e.g. $\rho_0 = 2.0$ and $\rho_1 = 0.5$ where ρ = utilization factor = λ / μ).

The service capacity control of a queuing system where the server capacity is incremented in units of basic bandwidth units (BBUs)—e.g. B-channel units of 64 kbps—and which is limited only by the upper allowable bandwidth limit (UPPER_BW_LIMIT), is a logical extension of the above models. These types of control policies are called *multi-level hysteretic service rate control* policies. I have found few results in the literature for this type of control policy for the M/M/1 queuing system with variable capacity in which the control policy implemented is of a hysteresis type. The model studied by Moder and Phillips (1962) is the closest to this study, and it considers an M/M/*c* type queuing system with identical parallel servers operating under a hysteresis threshold type of control. The main assumptions in that study are the instantaneous switching times and limited number of servers (see Section 1.6 above).

[3] The mean service rate (completion rate), $\mu = C/S$, where C is the server capacity (service units/s) and S is the mean work demand of a customer (service units). Note that S/C is the mean service time (s).

Here, it is assumed that the user–network interface is of the PRISDN type and is composed of a 30 B + D-channel structure. (Note that these policies apply to any $n \times$ B or $n \times 64$ kbps interface where $n \geq 2$.) We further assume that the increase/decrease in channel capacity is in units of one B-channel at a time, as it must be, given the current capabilities of ISDNs. The following four basic heuristic policies are defined (presented in a syntax similar to 'C'):[4]

- *Policy 1 (P1)* Increase and decrease BW *linearly* as follows:
```
Increase BW: if( (qsize >= TH_HIGH) AND (!pending) AND (bw < UPPER_BW_LIMIT) )
                    CHBW( 'I', 1 )
Decrease BW: if( (qsize == TH_LOW) AND (!pending) AND (bw > 0) )
                    CHBW( 'D', 1 )
```
- *Policy 2 (P2)* Same as policy 1 except the last channel (BBU) is always kept:
```
Increase BW: same as P1.
Decrease BW: if( (qsize == TH_LOW) AND (!pending) AND (bw > 1) )
                    CHBW( 'D', 1 )
```
- *Policy 3 (P3)*
  Increase BW: same as P1.
  Decrease BW: if the bandwidth is greater than 1 BBU, remove all channels but one; if the BW is 1 BBU, remove it.
```
        if( (qsize == TH_LOW) AND (!pending) )
        {
                if (bw > 1 ) CHBW( 'D', BW-1)
                else if( bw == 1 ) CHBW( 'D', 1 )
        }
```
- *Policy 4 (P4)*
  Increase BW: same as P1.
  Decrease BW: halve the existing capacity at every request.
```
        if( (qsize == TH_LOW) AND (!pending) )
        {
                if( bw > 1 ) CHBW( 'D', BW/2)
                else if( bw = = 1 ) CHBW( 'D', 1 )
        }
```

It is known that the optimal epochs for a change in the service rate of a queuing system is at the epochs of change in the queue size (Zachs and Yadin 1970). Hence the test for increasing the BW of a superchannel is initiated *after* queuing each arriving packet and updating the

[4] *qsize*: variable holding instantaneous (or averaged/weighted) queue size; could also be the *npkts* (number of packets in the system)
bw: variable holding the current bandwidth of a given superchannel
pending: INC/DEC of bandwidth is currently pending (! negates logic)
TH_HIGH: threshold high (also TH_H or H)
TH_LOW: threshold low (also TH_L or L)
CHBW: change bandwidth function with parameters (direction, bandwidth)
INC/DEC: bandwidth increase/decrease ('I'/'D') functions, respectively

qsize. Similarly, the test for decreasing the BW is initiated only after the reception of a 'packet-transmitted' message from the ISDN interface (returned as a PKT_SENT signal) which updates the *qsize*. Note that a positive feedback condition is created if the bandwidth increase operation is additionally done on the PKT_SENT message when the *qlen* is decremented. This may indeed be necessary for a responsive control.

In all the policies mentioned above, the method used for increasing the bandwidth is the same. The operation is as follows. When the very first packet arrives for a given destination, and there is no existing channel to the far end, a SETUP is initiated by the channel management software. It calls a function *open_cct()*, which creates an internal message block structure with the SETUP message and forwards it to the D-channel process (DCP signalling software) at the user–network interface (see Chapter 4 and Deniz and Knight 1989b). All further packet arrivals to the same destination are queued to the reserved channel queue while the previous channel set-up is still pending. Once the channel is set up and transmission starts, the channel bandwidth increase/decrease is made possible. Whenever the *qsize* reaches the threshold high (TH_HIGH) from below, or when a queue size update at a packet arrival finds the *qsize* at or greater than the TH-HIGH value, the channel bandwidth increase is initiated. While a bandwidth increase (or decrease) operation is ongoing (i.e. a change pending), further changes are inhibited. In the simulation model implemented, the channel switch-on (set-up) operation is assumed to take a time that is exponentially distributed with mean switching time (SW_TIME) γ. A BBU (i.e. time slot) disconnect is assumed to take effect immediately (zero switch-off time).

The above-mentioned operational characteristics mean that, as the queue size of the channel grows beyond the TH_HIGH, every new arrival could potentially initiate a new channel set-up (and aggregation) if the channel increase is not inhibited. This class of policies has a *greedy* characteristic such that, unless an upper limit is imposed on the bandwidth of the superchannel thus formed, it could (momentarily) possess all the available bandwidth at the interface, depending on the arrival rate. However, interestingly, the mean overall bandwidth used will always be just as much as required by the work-load. This presents certain problems regarding performance which are discussed in the latter sections. In a similar fashion, every time the queue size falls to the threshold low (TH_LOW) value from above, or when a 'packet transmitted' message encounters a *qsize* less than or equal to TH_LOW, the channel bandwidth is reduced still further. This type of operation has a tendency to over-compensate for changes in the queuing system behaviour and is discussed below.

The main difference between these four heuristic policies is in the way in which the bandwidth is reduced by the TH_LOW triggering. Policy P1 releases a channel every time the 'packet transmitted' message finds the system at or below the TH_LOW. As this operation is assumed to take zero time, the rate at which the channel bandwidth decrease can be achieved is potentially faster than the increase operation. Policy P2 always leaves one last channel open. The rationale behind this policy is that it would be always ready for further packet arrivals and hence would incur less cost arising from time delay in setting up a new channel. A mechanism for closing this one down could be linked to a time-out mechanism where the absence of activity over the channel for a period given by the time-out could trigger a complete dismantling of the channel. In policy P3 all but one channel of the superchannel are removed. If the condition persists, the last channel is also removed. In policy P4, the existing bandwidth is always halved when the 'packet transmitted' signal causes a bandwidth decrease trigger.

A natural extension of the policies is the limiting of bandwidth increase/decrease operations to the crossing of the levels from *below* or *above*. This should reduce the number of switchings per packet (or unit time) as well as the maximum amount of bandwidth assigned to a given superchannel for a given packet arrival rate. However, this type of policy necessitates the use of multiple threshold values for both the increase and decrease bandwidth operations, since under overload conditions the *qsize* may never fall below the TH_HIGH value or, under extreme underload conditions, go above the TH_LOW value of the above models.

8.6 ANALYSIS OF PERFORMANCES

These are very reactive (and greedy) policies. They can be modified to smooth their reactivity, giving a less responsive system but more efficient management of the bandwidth resource.

The performance of these policies is shown in Figs 8.2–8.14. All of the results have been obtained by running models with the policies mentioned previously and using a maximum allowable channel capacity of 30 BBUs (30 x 64 kbps). In each case the following assumptions are implemented. The packet sources are of Poisson type, giving packet inter-arrival times (IATs) that are exponentially distributed with mean $1/\lambda$. The packet size distribution is again assumed to be exponential with mean PSIZE (1000 bytes/packet). At this packet size setting, each BBU (B-channel) has a mean service rate μ=8 packets/second (pps). Each run of the simulator was conducted with 100 000 packets, and an overhead of 10 per cent was added for the warmup effect (hence a total of 110 000 packets/run). Each point on the graph is obtained from a replication of 10 runs.

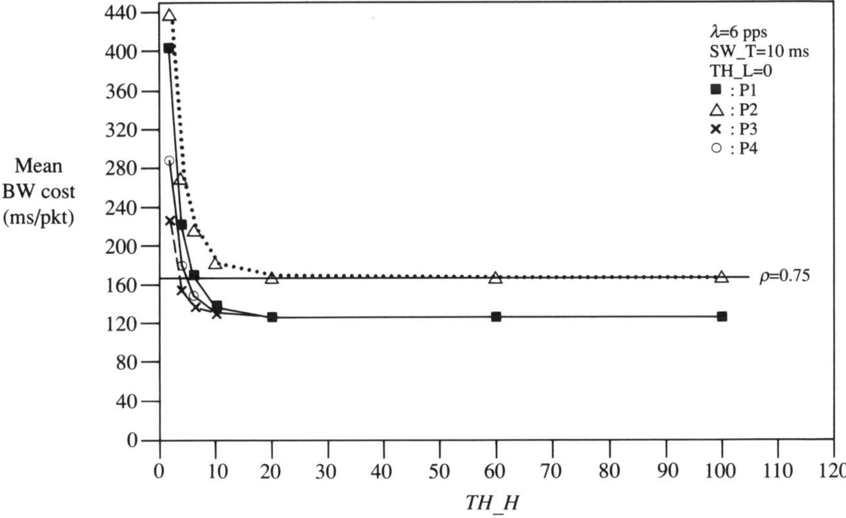

Figure 8.2 Mean bandwidth cost *v*. threshold high (λ=6 pps)

Figures 8.2–8.4 shows the mean bandwidth cost (C_{BW} in BBU ms/packet) versus the threshold high setting (TH_H). In all the experiments, the low threshold has been kept at zero (TH_L = 0). The mean bandwidth cost represents the time duration that the channel was kept open on a per-packet base. The actual cost is given by (mean bandwidth cost x BBU cost/ms). The values for bandwidth costs have been measured over the whole duration of the experiment and the usage of multiple BBUs has been accounted for. In Fig. 8.2, the arrival rate parameter λ is set to 6 pps. This means that, in the case of non-switchable channels, one channel at 64 kbps must be provided for the duration necessary to transmit 110 000 packets. This can be assumed to be (110 000 x 1/6) / 110 000/packet (= 166.67 ms/packet). One BBU at 8 pps will operate at a traffic intensity of $\rho = \lambda/\mu = 6/8 = 0.75$. This line is shown in Fig. 8.2. Similar lines are drawn in Figs. 8.3 and 8.4. These correspond to 7 and 10 BBU usages, respectively. The curves above this line show the trade-off between the bandwidth (holding) cost and the response time. Below this line, the holding cost is less than that of the non-switchable channel but at the cost of increased response times. As the packet arrival rate parameter is increased, the distinction between the policies P1 and P2 diminishes. This is as expected, since the mean bandwidth used is very much larger than 1 BBU and the keeping of the last channel in P2 does not make much difference on the performance. Furthermore, P2 tends towards the limit of one non-switched channel performance for values of TH_H greater than 20 packets, as would be expected. Figures 8.2–8.4 also show that, in terms of the bandwidth costs incurred, policy P3 performs best at low arrival rates while policy P4 performs better at high values of λ. This is because in P3 closing all but one of the channels while the arrival rate is high has a detrimental effect, since more channels must then be opened up quickly to satisfy the need of the traffic. Policy P4 presents a compromise, in that it halves the bandwidth capacity when the low threshold is triggered, rather than taking all channels out of service.

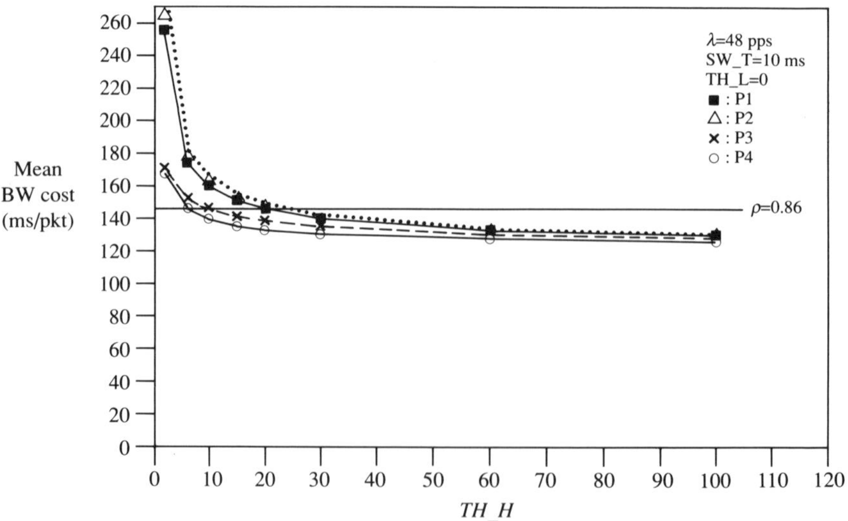

Figure 8.3 Mean bandwidth cost *v*. threshold high (λ=48 pps)

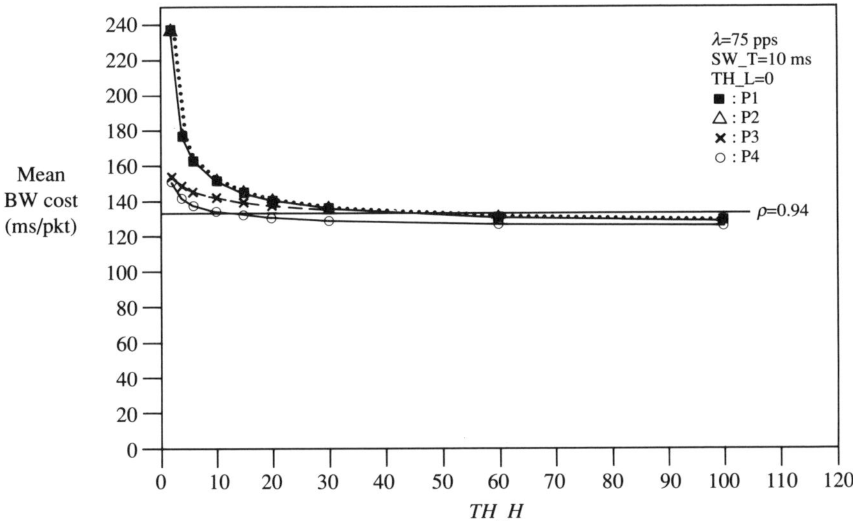

Figure 8.4 Mean bandwidth cost *v*. threshold high (λ=75 pps)

Figures 8.5–8.7 show the effect of increasing the mean switching time, γ. Comparing Figs 8.2, 8.5 and 8.6, it can be seen that at low packet arrival rates the bandwidth holding cost in policy 2 is always higher than the rest, as would be expected, since the last channel is always kept on while the other policies switch it off when it is not needed. The difference between policies P1, P3 and P4 also diminishes as the mean switching time is increased. Comparing Figs 8.4 and 8.7, we can see that the difference between policies P3 and P4 diminishes while that between P1 and P2 increases, with P2 approaching the P3 and P4 performances. It is interesting to note that in Fig. 8.7 policy P2 performs better in terms of bandwidth costs than 10 fixed channels (ρ=0.94) for TH_H values of 12 packets or greater. This is expected, as keeping 1 channel in 10 (on average) all the time does not make much difference to the performance. However, the comparative performance of P1 and P2 changes drastically, with large switching times at high arrival rates (see Figs 8.4 and 8.7). This is explained by the fact that any 'error' in bandwidth is over-compensated by P1 while P2 relies on the one channel always left open to prevent over-compensation. In the case of P3 and P4, the compensation is forced at an earlier bandwidth level since they shut down 'extra' bandwidth more drastically then either P1 or P2, causing even less over-compensation than P2.

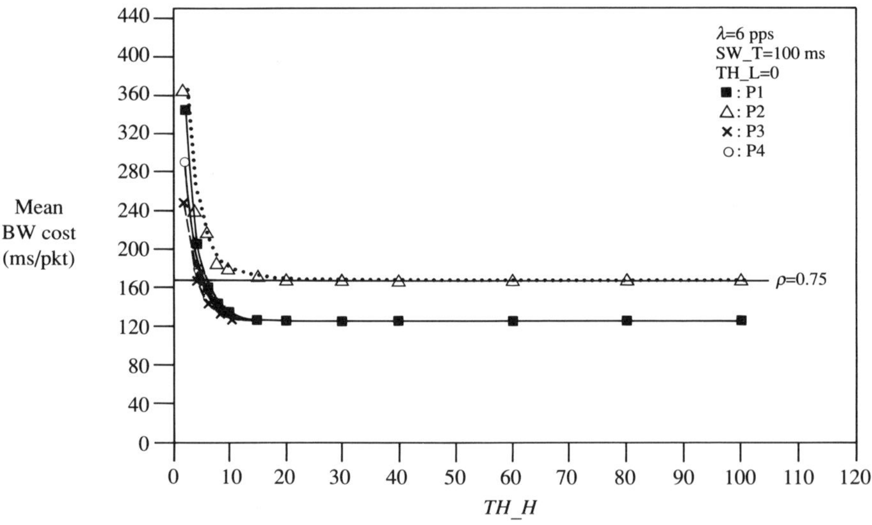

Figure 8.5 Mean bandwidth cost *v*. threshold high (λ=6 pps; SW_T=100 ms)

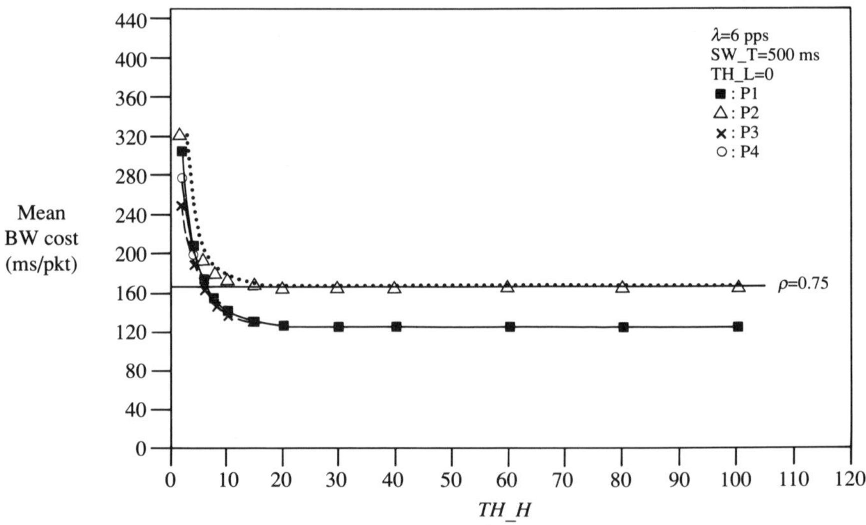

Figure 8.6 Mean bandwidth cost *v*. threshold high (λ=6 pps; SW_T=500 ms)

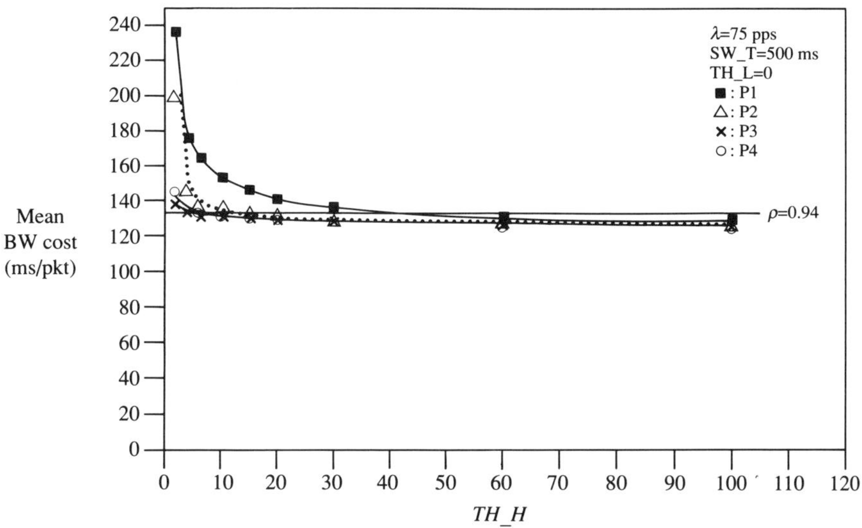

Figure 8.7 Mean bandwidth cost *v.* threshold high (λ=75 pps; SW_T=500 ms)

Figures 8.8 and 8.9 show the mean response time, *RT*, for the four policies at two threshold settings, i.e. TH_H = 10 and 100, while TH_L = 0 in both cases. Policy P3 gives the highest values in both cases, while P4 provides a compromise between the two extremes given by P3 versus P1 and P2. The difference between the policies disappears as the threshold high is increased. The behaviour of the curves is interesting, in that it rises until a value of $\lambda = 8$ pps is reached (corresponding to $\rho = 1.0$), and then falls almost as $1/x$. This is expected if the curves of Figs 8.10 and 8.11 are observed. Figures 8.10 and 8.11 show the variation of the mean number of customers (packets) in the system (*N*) with the packet arrival rate λ. All the policies show a definite turn in their rate of increase beyond $\lambda = 8$ pps. After the turning point, policies P3 and P4 show a linear increase with gradients larger than those of P1 and P2. In fact, P1 and P2 hardly increase their value beyond $\lambda = 8$ pps ($\rho = 1.0$), From Little's result, the RT is given by N/λ. Hence, for *N* relatively constant with increasing λ, this gives the '$1/x$' type behaviour beyond the critical point of $\rho = 1.0$.

Figures 8.12–8.14 show the mean number of switchings per packet over the range of high thresholds: TH_H=2, 10 and 100. Policy P2 provides the least number of switchings below the arrival rate given by $\lambda = 8$ pps. This is expected, since it always provides one channel. Beyond this turning point, it performs virtually identically to P1. At high threshold values P3 always performs worse in that it achieves the highest number of switchings per packet. However, as shown in Fig. 8.12, policies P3 and P4 perform better beyond a second turning point at approximately $\lambda = 50$ pps.

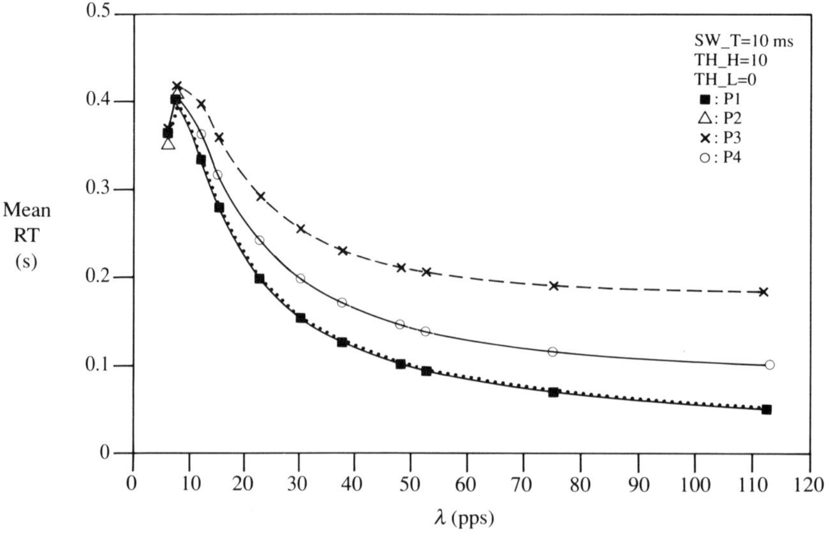

Figure 8.8 Response time *v*. packet arrival rate (TH_H=10, TH_L=0)

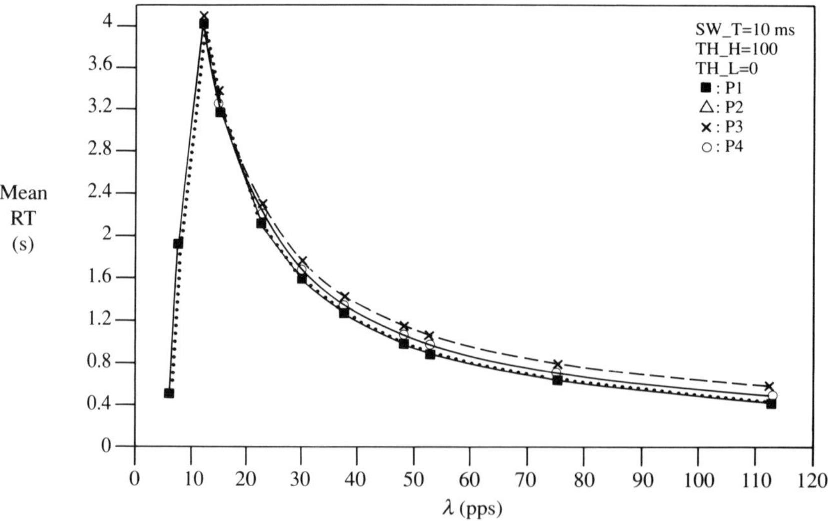

Figure 8.9 Response time *v*. packet arrival rate (TH_H=10, TH_L=0)

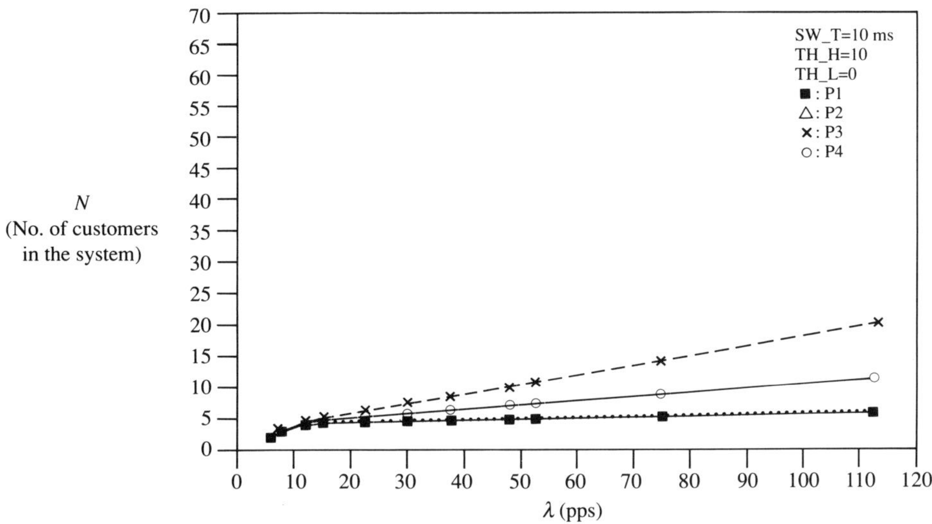

Figure 8.10 Mean number of customers *v*. packet arrival rate (TH_H=10, TH_L=0)

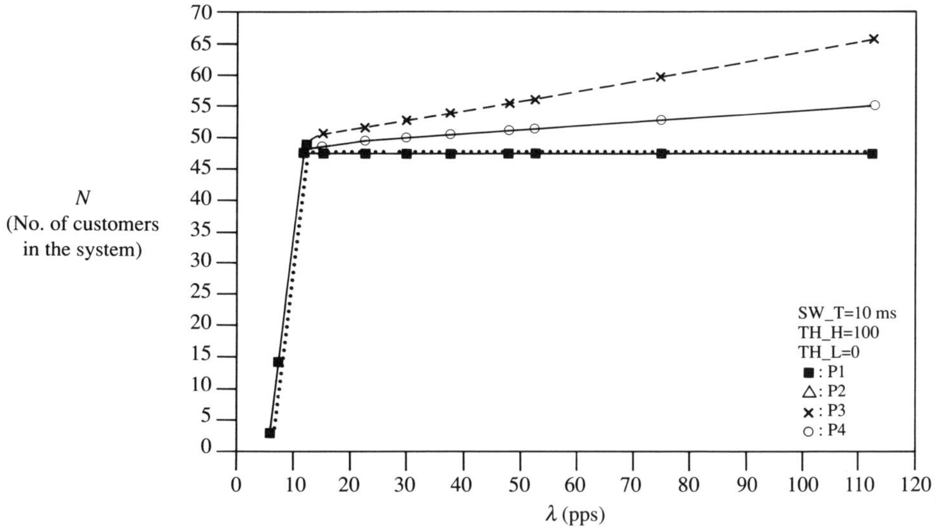

Figure 8.11 Mean number of customers *v*. packet arrival rate (TH_H=100, TH_L=0)

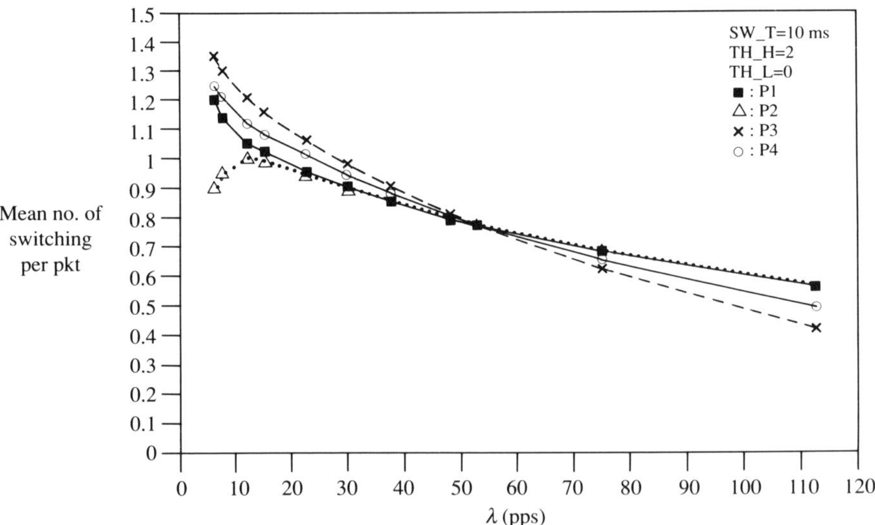

Figure 8.12 Mean switching *v*. packet arrival rate (TH_H=2, L=0)

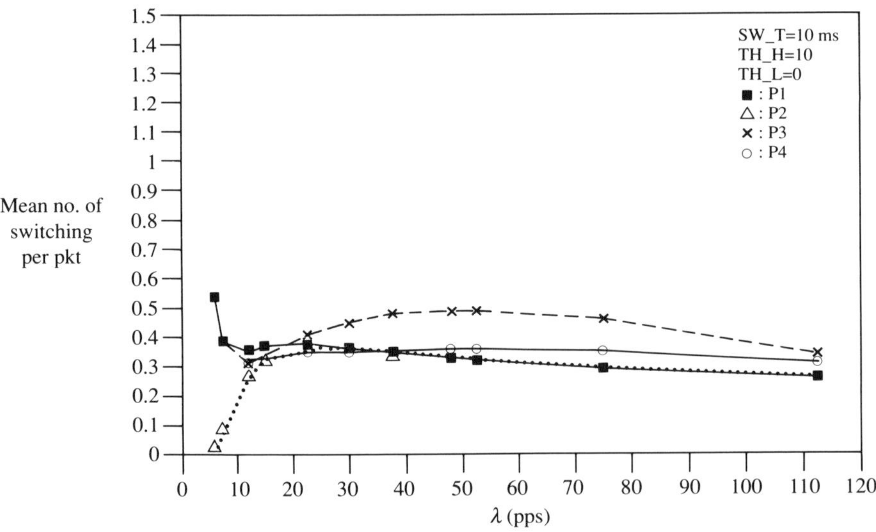

Figure 8.13 Mean switching *v*. packet arrival rate (TH_H=10, L=0)

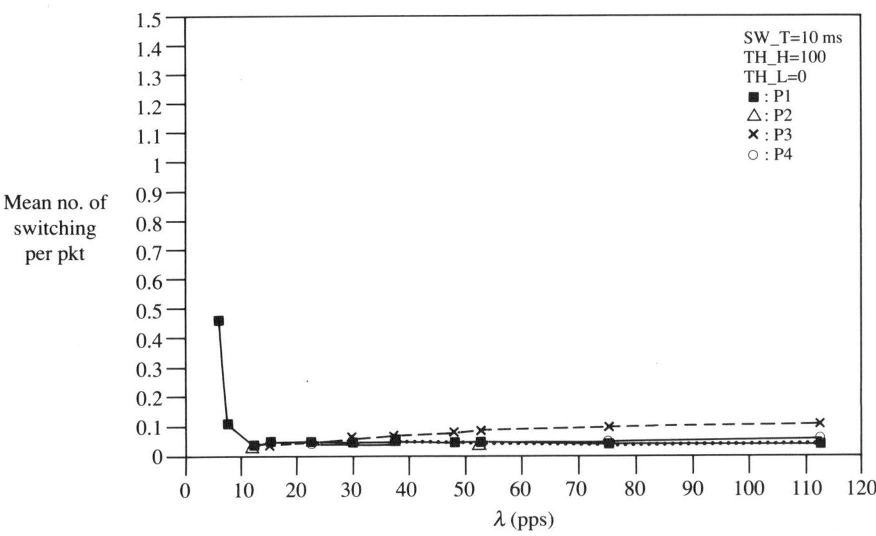

Figure 8.14 Mean switching *v*. packet arrival rate (TH_H=100, L=0)

8.7 CONCLUSIONS

Figures 8.2–8.7 show that the value of bandwidth cost (and hence the average bandwidth used) can be increased or decreased by varying the threshold high value while keeping the threshold low value at zero. This has the major implication that the cost–performance trade-off can be controlled by the use of this class of policies. The effects of increased arrival rate are threefold:

1. The difference between the maxima and minima of the curves is reduced, leading to more gradual response curves.
2. The difference between policies P1 and P2 disappears and that of P3 and P4 reverses.
3. The *critical threshold* value (that above which the bandwidth cost is lower and below which the bandwidth cost is higher) increases.

Another observation is the fact that, as the corresponding traffic intensity, ρ, increases, the policy curves provide less savings in the bandwidth cost beyond the critical threshold value. That is, the fixed-channel case provides a limiting value for the heuristic policy curves.

The effect of increasing the mean packet arrival rate, λ, at constant mean switching time, γ, is to reduce the difference between the policy performances. The relative performances of P3 and P4 are reversed. The effect of increasing γ while keeping λ constant is similar, except the relative performances of P3 and P4 stay the same.

8.7.1 Sensitivity of performance

These are very reactive (and greedy) policies in that they can momentarily grab large amounts of bandwidth if allowed. They can be modified to smooth their reactivity, giving a less responsive system but more efficient management of the bandwidth resource. Two methods of achieving this are shown in Chapter 9: the 'moratorium' and 'delayed close' policies.

8.7.2 Characterization of queuing system performance

It can be seen from the results that this class of policies tends to use the bandwidth in such a way that the average bandwidth utilization is a minimum for a given utilization factor ρ. The bandwidth utilization of the system can be increased (and hence the average response time of the packets reduced) if the queuing system is forced to switch-in more channels by the reduction of the high-threshold (TH_H) value while keeping the low-threshold (TH_L) value constantly at zero. This produces the undesirable effect of increased queue sizes (and hence a *peaked* response time performance, shown in Figs 8.8 and 8.9) when the system operates close to $\rho = 1.0$, i.e. when the packet arrival rate is close to the capacity of 1 BBU.

We conjecture that this critical value would still exist if the minimum number of BBUs were increased to higher values (e.g. 2, 3, etc.). The only effect would be to increase the critical point to corresponding values of ρ (e.g. $\rho = 2, 3$, etc.). The characteristic behaviour of the queuing system would remain unchanged.

8.8 SUMMARY

In this chapter, four queue-based bandwidth management policies were proposed, and their performances evaluated and compared using a specially developed simulation model. This model includes two CL–NL relays interconnected via an ISDN.

Channel and bandwidth management constitute a very important and worthwhile area necessitating further studies; they affect the performance, efficient resource utilization and operational costs of data communications using the ISDN, especially in the multiplexed user traffic and multi-media application environments.

CHAPTER

NINE

COMPARISON OF QUEUE-BASED SERVICE CAPACITY CONTROL DOCTRINES

Service capacity control doctrines based on the threshold principle can be classified into different categories according to the *form* of the threshold policy, the *number of control levels* for the service capacity and the *number of thresholds used*. The form of the threshold policy can be either a simple (single or point) threshold or one using a hysteresis form. The number of control levels for the service capacity can be either uni-level, bi-level or multi-level. In uni-level control the queuing system server(s) are switched on as a threshold is crossed and switched off when the same or another threshold is crossed in the reverse direction. This is an *on/off* type of control (see Yadin and Naor 1963). In bi-level control, the service capacity of the queuing system oscillates between two distinct service capacities. This method is sometimes known as an *overload/underload* type of control, since the lower service capacity is usually less than the arrival rate and the upper service capacity is greater than the arrival rate of the traffic under control. In multi-level control, the service capacity is allowed to have a number n ($n > 2$) of integral bandwidth values between lower and upper service capacities. The number of thresholds, or threshold pairs (in the case of the hysteretic form), used can be single or multiple. In the case of multiple hysteretic thresholds, these pairs can be overlapping or non-overlapping in their placement along the measure index used, and they could have uniform or non-uniform hysteresis loops.

The metric used as the measure index in threshold control can be one of several variables. The most common are the instantaneous or averaged values of the queue length or the customer arrival rate; variables such as the number of customers in the system can also be used. In this book it is mainly the latter that has been used, although various measurements have also been taken with the other measure indices.

In Chapter 8, four heuristic multi-level hysteresis (single-pair) threshold control policies have been described. In this chapter, other threshold policies are evaluated and compared with these four. Some variations on the theme of multi-level hysteresis threshold control are also presented; these are the 'delayed close' and the 'moratorium' policies.

9.1 BI-LEVEL SERVICE CAPACITY CONTROL

Bi-level service capacity[1] control using thresholds over the queue length form a class of control policy where the service capacity of the system under control can have one of two distinct service capacities, given by a lower bandwidth (BW_LOW) and higher bandwidth (BW_HIGH). The service capacities give the lower and upper mean service rates, denoted by ρ_0 and ρ_1. These control policies, applied to an M/M/1 queuing system with variable service rates, have been studied by Yadin and Naor (1963), Gebhard (1967) and later by Harita and Leslie (1989). The model by Yadin and Naor is for a general M/M/1 queue whose service capacities can be increased or decreased in discrete steps according to the instantaneous queue size. They generalize the results for a queuing system with multiple hysteresis loops whose hysteresis width can be varied (collapsing to a point threshold in the limit), which can have non-uniform hysteresis widths and overlapping or non-overlapping threshold pairs. An expression for the expected queue size is obtained in closed form. However, this expression is rather cumbersome for solution and is not used here. The other two models are studied in the following two sections.

9.1.1 Bi-level capacity control with instantaneous switching

The model studied by Gebhard (1967) is an M/M/1 queuing system whose channel capacity is variable according to a threshold doctrine imposed over the queue length. This study includes the steady-state solutions for this system when operating under point and hysteresis threshold policies. Closed-form expressions for the mean number in the system and the mean queue length are obtained. Gebhard's study also includes two measures of effectiveness and a cost analysis under different cost structures.

However, Gebhard presented no graphical solutions to the equations he obtained. Here, the equations provided for hysteresis control are solved and graphically depicted for ease of understanding of their characteristic behaviour.

Equation (9.1) shows the closed-form solution obtained by Gebhard for the mean number in the system, N, for a queue under bi-level hysteresis (threshold) service rate control:

$$N = P_0 \left\{ \frac{\rho_0}{(1-\rho_0)^2} - \frac{(p+1)\,\rho_0^{L+p}\,(\rho_0-\rho_1)}{(1-\rho_0^{p+1})\,(1-\rho_1)} \left[\frac{2L+p}{2} + \frac{(1-\rho_0\rho_1)}{(1-\rho_0)\,(1-\rho_1)} \right] \right\} \tag{9.1}$$

where $p = H-L-1$, H is the higher threshold (upper control level) and L is the lower threshold (lower control level) in terms of the number of customers in the system. Also, ρ_0 and ρ_1 are the traffic intensities at the lower and higher service rates; i.e., $\rho_0 = \lambda/\mu_0$ and $\rho_1 = \lambda/\mu_1$.

Under this type of service capacity control, the service rate of the server is increased from μ_0 to μ_1 ($\mu_1 > \mu_0$) whenever the queue length increases from below a value to a value equal to or greater than H ($H > L$). The service rate is reduced back to μ_0 only when the queue length falls to a value equal to or less than L.

[1] Note that the service (server) capacity, the channel capacity and the channel bandwidth are used synonymously in this book. A service rate results from a given server capacity and customer (packet) work load (see Chapter 8).

Equation (9.2) shows the solution to the variable P_0:

$$\frac{1}{P_0} = \frac{1}{(1-\rho_0)} - \frac{(p+1)\,\rho_0^{L+p}\,(\rho_0-\rho_1)}{(1-\rho_0^{p+1})\,(1-\rho_1)} \qquad (9.2)$$

Figures 9.1–9.10 show the behaviour of this overload/underload system under different load conditions. Figures 9.1 and 9.2 are plotted for a hysteresis model where the high threshold (TH_H, or H) is set to 6 (customers or units in the queue) and the low threshold (TH_L, or H) is set to zero. Figure 9.1 shows the variation of mean queue size (Q) with ρ_1. Different curves are plotted for various values of ρ_0. Two important conclusions can be drawn from these graphs. First, as the value of ρ_0 grows beyond 1, the curves come nearer to each other for values of $\rho_0 > 1$. Therefore, at high traffic overload values ($\rho_0 >> 1$), the important metric in the control becomes the value of ρ_1. At low load values ($\rho_0 < 1$), both the ρ_0 and ρ_1 values are of importance in determining the system behaviour. Second, the operating value of ρ_1 is not very critical between the values of 0–0.7 (approximately) in terms of the values of Q or the average system response time (RT), but they both go to infinity as the higher bandwidth reaches its control limit nearer the value of $\rho_1 = 1$.

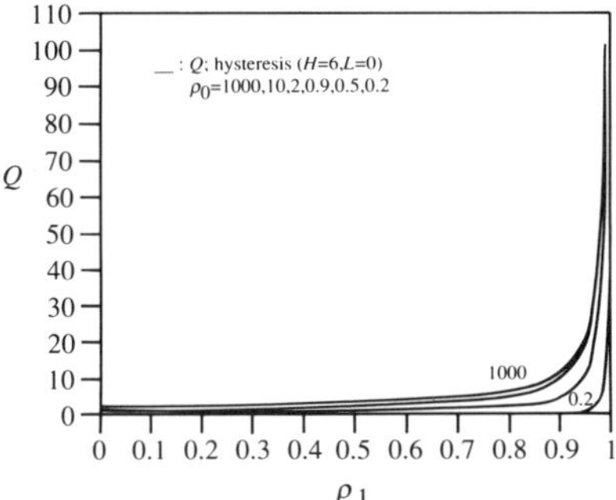

Figure 9.1 Mean queue length *v.* ρ_1 (Gebhard H=6, L=0 model)

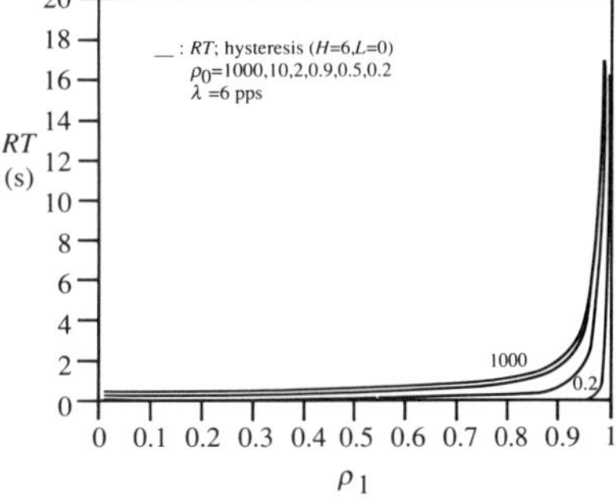

Figure 9.2 Mean response time *v.* ρ_1 (Gebhard H=6, L=0 model)

Figures 9.3 and 9.4 show the variation of Q and RT versus ρ_1 for different ρ_0 values for a bi-level hysteresis control system where the higher and lower threshold values are set to 100 and 50, respectively. From Fig. 9.3 it can be seen that the mean queue length of the controlled system is established between the lower and upper threshold values (50–100).

An important feature of this system is demonstrated in Figs 9.3 and 9.4, in that the curves are very flat for values of ρ_1 between 0 and 0.9. This leads us to the conclusion that the operating value of ρ_1 must be determined by cost factors such as number of switchings per second, since a ρ_1 value of 0.5 provides the same mean response time (for the same arrival rate) and mean queue size as the value of 0.1 (see Fig. 9.4).

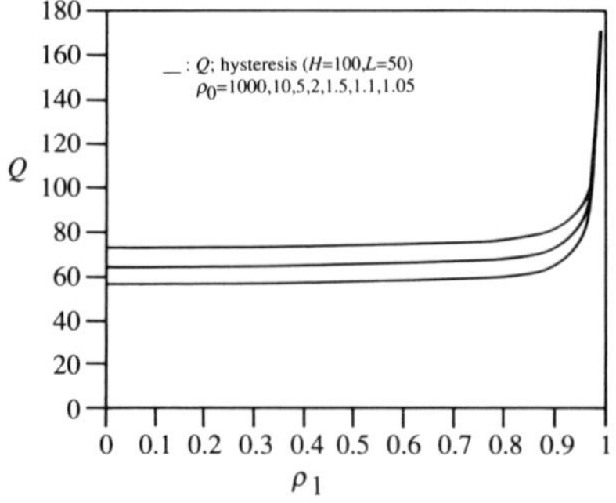

Figure 9.3 Mean queue length *v.* ρ_1 (Gebhard *H*=100, *L*=50 model)

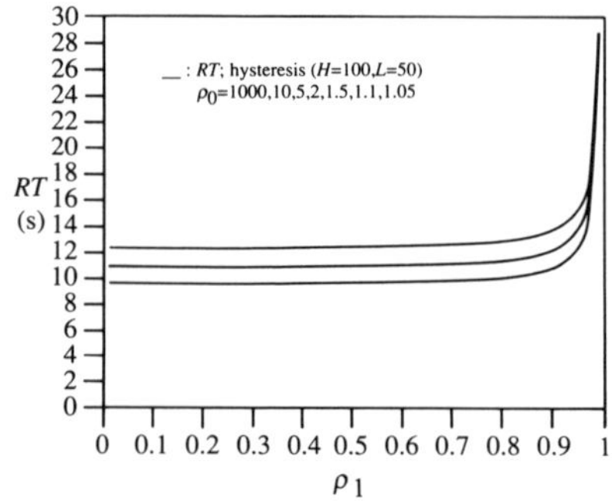

Figure 9.4 Mean response time *v.* ρ_1 (Gebhard *H*=100, *L*=50 model)

Figure 9.5 shows the plot of mean response time in seconds versus the mean arrival rate, λ, in packets per second. The curves that are plotted for different ρ_0 values and a fixed $\rho_1 = 0.5$ value are found to lie close to each other. Figure 9.6 shows a plot of the mean number of switchings per second versus ρ_0 for different settings of ρ_1 (0.1–0.9). This graph shows that the critical values of ρ_0 are those between 0 and 10. This is in agreement with Figs 9.1–9.5 in that, for values greater than $\rho_0 > 10$, the system provides closely fitting performances. In other words, once the system is overloaded, it does not matter whether it is 10, 100 or 1000 times overloaded as long as the higher bandwidth is able to cope with the traffic. In reality, all it determines is how quickly the upper bandwidth is switched in and for how long it is kept in service.

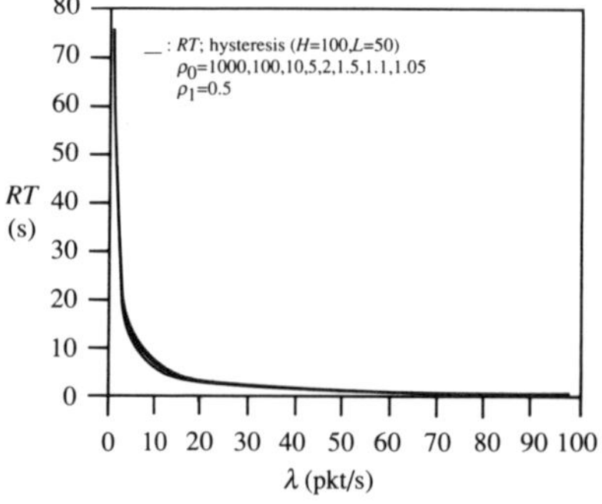

Figure 9.5 Mean response time *v.* λ (Gebhard *H*=100, *L*=50 model)

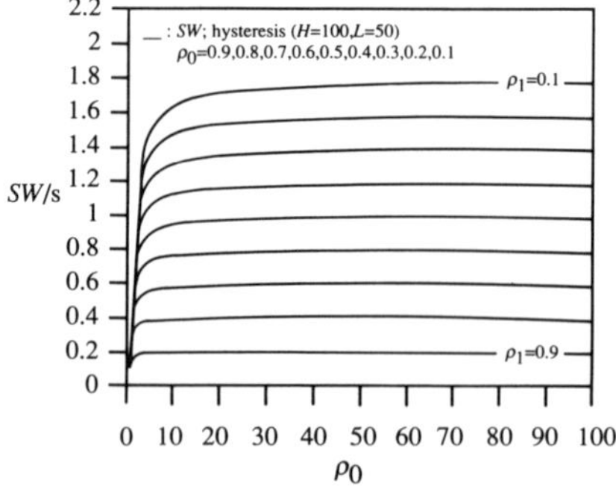

Figure 9.6 Mean number of switchings per second *v.* ρ_0

Figure 9.7 shows the variation of the mean number of switchings (*SW*) per second versus ρ_1 for different ρ_0 settings. The variation of *SW* with ρ_1 becomes noticeable for values of $\rho_0 > 1.5$. Also, for values of $\rho_0 > 10$ the curves obtained give very similar performance values for *SW* against ρ_1. Figure 9.8 shows the variation of *SW* versus the packet arrival rate, λ (packets/second). The trends observed for Fig. 9.7 also apply here.

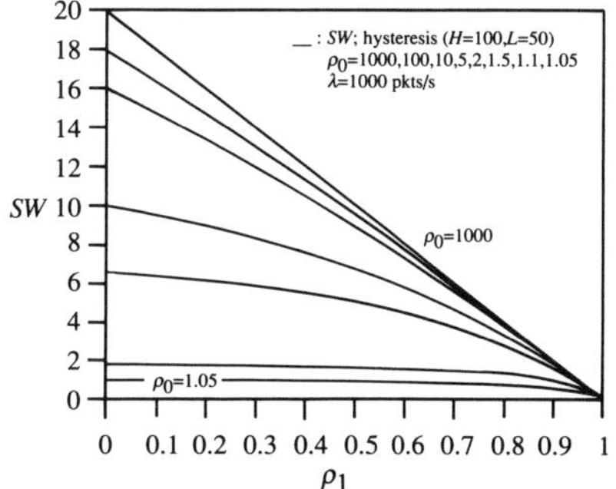

Figure 9.7 Mean number of switchings per second *v*. ρ_1 (Gebhard *H*=100, *L*=50 model)

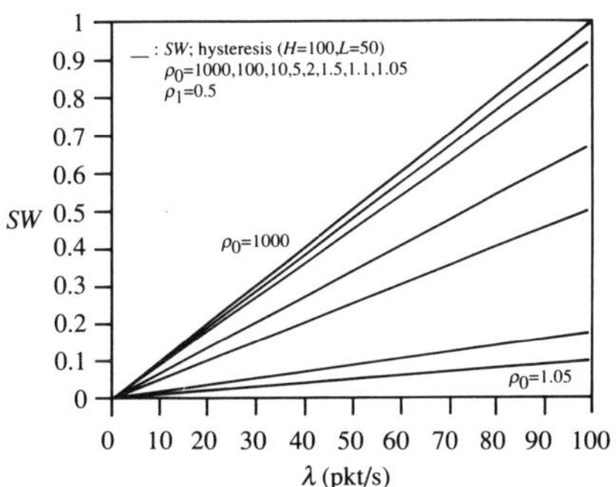

Figure 9.8 Mean number of switchings per second *v*. λ

Figure 9.9 shows the variation of *SW* versus the lower threshold (L or TH_L)[2] for different values of ρ_0 where $\rho_1 = 0.5$ and TH_H=100. As expected, the number of switchings increases drastically as the lower threshold approaches the upper threshold value. Figure 9.10 shows the variation of *SW* versus the difference between the low and high thresholds, given by variable *P*. This is more or less the mirror image of Fig. 9.9.

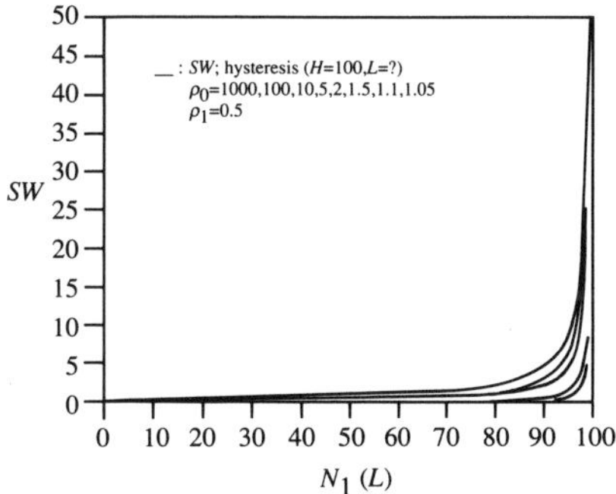

Figure 9.9 Mean number of switchings per second *v*. N_1 (Gebhard *H*=100, *L*=50 model)

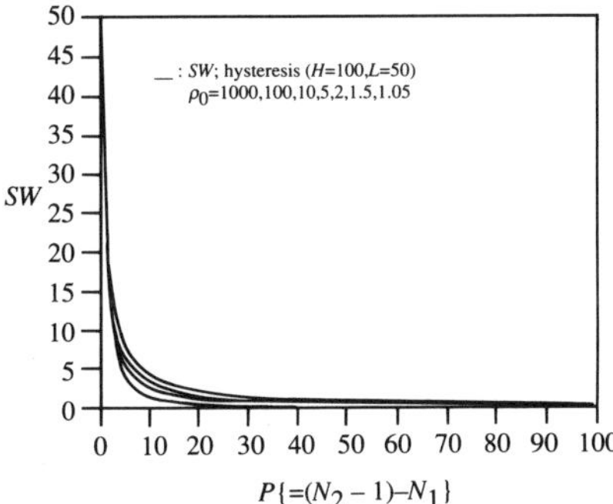

Figure 9.10 Mean number of switchings per second *v*. *P*

9.1.2 Bi-level service capacity control with switching delays

The models studied by Harita and Leslie (1989) include the point and hysteresis threshold forms of control policy operating under bi-level service capacity control of an M/M/1 queuing system with switching delays. The channel switching-on delay is assumed to be exponentially distributed with parameter γ and the switching-off delay is assumed to be zero. Closed-form expressions for the mean number of customers in the queue are derived for both cases.

[2] Note that in Gebhard's terminology $N_1 = L$ and $N_2 = H$.

A simulation model of the M/M/1 queue under the above-mentioned service capacity control policy has been implemented and its performance under different conditions evaluated. Table 9.1 shows the performance of the bi-level and multi-level hysteresis threshold control policies operating under similar constraints. It is assumed that the interface is composed of *n* x B-channels. It can be seen that the multi-level control policy performs better than its bi-level counterpart in having a better response time at a reduced number of switchings per packet and approximately the same bandwidth holding cost.

Table 9.1 Comparison of multi-level and bi-level service capacity control policies[a]

	Multi-level P2 (UL=4)[c]		Bi-level ($\rho_0 = 2 - \rho_1 = 0.5$)	
Measure[b]	H=10; L=0	H=100; L=0	H=10; L=0	H=100; L=0
N	5.348	48.42	5.744	50.394
RT (ms)	333.97	3022.45	359.85	3157.86
SW	0.141	0.0152	0.194	0.0199
BW (BBUs)	2.277	2.031	2.249	2.018
C_{BW} (ms/pkt)	142.19	126.91	140.82	126.53

[a] Both cases: $\lambda = 16$ pps, $\mu = 8$ pps, switching time (SW_T) = 10 ms.
[b] N: number of customers in system; RT: response time; SW: number of switchings/packet; BW: mean bandwidth used; C_{BW}: cost of bandwidth.
[c] UL: bandwidth upper limit.

9.2 MULTI-LEVEL SERVICE CAPACITY CONTROL

Multi-level service capacity control using thresholds over the queue length form a class of control policy where the service capacity of the system under control can have any one of the service capacities (usually assumed to be discrete and integral) available in the range of a lower bandwidth (BW_LOW) and higher bandwidth (BW_HIGH). These control policies applied to an M/M/c queuing system with variable service channels have been studied by Moder and Phillips (1962). Chapter 8 above presented four different control policies based on the multi-level service capacity control doctrine applied to an M/M/1 system with switching delays. I am not aware of any previous study into this class of queuing system operating under such policies. In the following sections a closer look is taken into the multi-level service capacity control doctrines.

9.2.1 Multi-level capacity control with instantaneous switching

The Moder–Phillips (1962) model assumes a hysteresis threshold control over the service capacity of an M/M/*c* queuing system. Capacity increases are assumed to occur instantaneously by the switch-on of additional parallel identical servers (each with a mean service rate μ), once the instantaneous queue length reaches the upper threshold level. This process continues until an upper bound on the number of active servers reaches *S*. Capacity decrease occurs when the instantaneous queue length falls back to the lower threshold value. At this time, servers completing their service are switched off. This process continues until a lower limit σ on the active channels is reached.

Closed-form solutions of *N* (or *L*), the number of customers (packets or units) in the system, and *Q* (or L_q), the number of customers in the waiting line for the resulting queuing system, have been developed by Moder and Phillips (1962). It has been found that the equation developed therein for *N* is erroneous, in that it gives negative results for some settings of the variables. This expression has been re-evaluated in the present research using the system equations given in Moder and Phillips (1962) and has led to a modified equation for *N*. It is this modified version of the equation that is used in this chapter. A detailed analysis of the differences in the two formulae is given in Appendix IV. Figures 9.11–9.14 show the performance of this class of policies under various configurations that are obtained using equation (IV.2) of that Appendix.

Figures 9.11 and 9.12 show the variation of *N*, the number of customers in the system, versus ρ, the traffic intensity. The figures show plots for different high and low-threshold values, and the maximum (*S*) and minimum (σ) number of servers allowed. It can be seen that the two curves follow the same shape. This behaviour is characterized by the following features: the number in the system shows a fast increase at around the value of ρ corresponding to the channel capacity of one server; it then levels off with a small positive gradient until the queuing system is saturated at the maximum number of channels (servers) available.

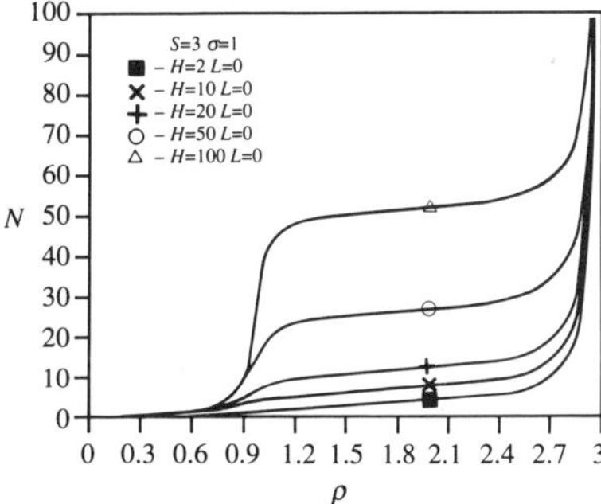

Figure 9.11 Mean number in system *v*. ρ (MP model S=3,σ=1)

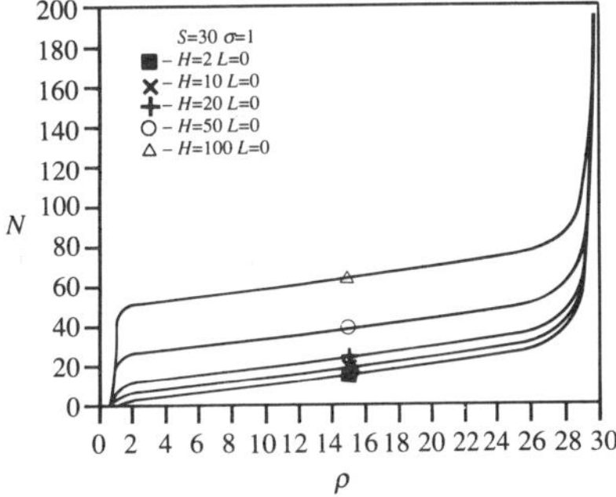

Figure 9.12 Mean number in system *v*. ρ (MP model S=30,σ=1)

Figures 9.13 and 9.14 show the variation of mean response time (*RT*) with the arrival rate for $\mu = 1$ packet per second. It can be seen that a peak in the response time occurs at around $\lambda = \mu$, the service rate of one server. This compares well with the results obtained from the simulation model for the (approximate) M/M/1 system in Chapter 8. Note that Figs 8.8 and 8.9 are obtained with $\mu = 8$ pps; hence a scaling factor of 8 is present in those curves.

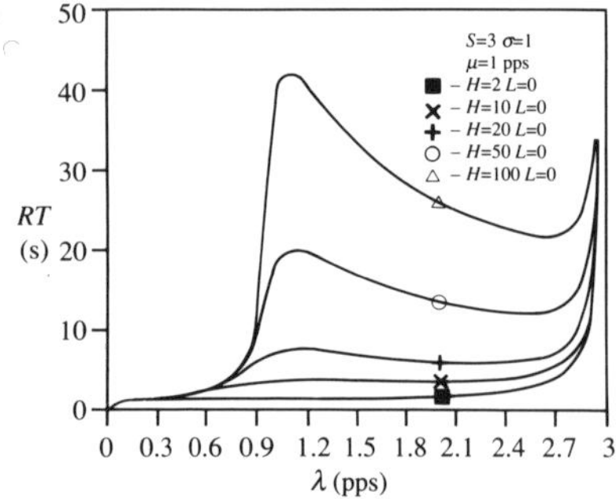

Figure 9.13 Mean response time *v*. λ (MP model S=3,σ=1)

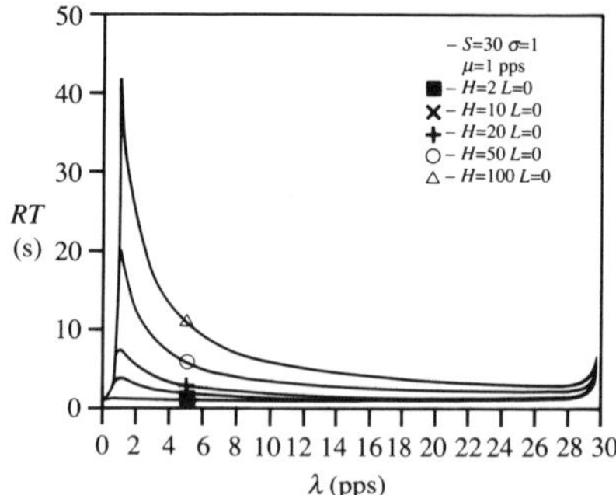

Figure 9.14 Mean response time *v*. λ (MP model S=30,σ=1)

9.2.2 Multi-level service capacity control with switching delays

I am not aware of any analytic study into this class of policies. Chapter 8 above presents four such policies for an M/M/1 queuing system operating under this class of control doctrine. Various complications have been added to the queuing system in order to make it as realistic as possible. The following section provides a comparison between the model studied in this book (with particular reference to Policy 2) and that studied by Moder and Phillips (1962).

9.2.3 Comparison of two multi-level capacity control systems

The Moder–Phillips (MP) model resembles in many respects *Policy 2* introduced in Chapter 8. However, several basic differences between the MP model and the models proposed in this book are recorded. The main ones are:

1. In the present model, the threshold values are built on the number of customers in the system (*N*), whereas in the MP model it is based on the number of customers in the queue (*Q*).
2. In the model proposed in this book, the channel switch-on time is assumed to be exponentially distributed with parameter γ, whereas the MP model assumes instantaneous switching times.
3. The present model assumes that any increase in the bandwidth of a channel is effective (following a switching delay) only after the completion of the current packet servicing to achieve channel synchronization. (However, the channel synchronization time is assumed to be zero.) The MP model assumes that a new server (channel) can be initiated instantaneously after the crossing of the high threshold without any notice of the state of other channels (as they form an M/M/*c* type queuing system).

4. According to the above comparison, the value of N in the present model is reduced only upon transmission of a packet at the new channel capacity; that is after the appropriate delays introduced by both the channel switching delay and the remaining transmission time of the packet previously in service are accounted for. In the MP model, the queue length Q is immediately reduced as a new channel is initiated (if a free channel exists). When the upper bandwidth limit is not yet reached and the queue length is one less than the upper threshold value, an arriving packet effectively initiates an idle channel into service immediately while the queue length stays the same.
5. Releasing of the bandwidth is different in the two systems. In the present model, as long as the number of customers in the system is zero, only one basic bandwidth unit is released; each further release of a BBU needs to have a packet arrival and a subsequent departure. In the MP model, as long as the queue length drops to the lower threshold setting (*qlen=TH_L*), all 'extra' bandwidth units (servers) are released as soon as each becomes idle.

The following similarities are noted. Both the M/M/1 and the M/M/c queuing systems give a characteristic behaviour in that the number of customers in the system increases dramatically at high values of upper-threshold settings when the mean arrival rate approaches the mean service rate of one channel (i.e. when $\rho = 1.0$). The characteristic behaviour of both types of queuing system is independent of the maximum number of channels used, S.

9.3 MODIFIED MULTI-LEVEL SERVICE CAPACITY CONTROL POLICIES

Two simple modifications can be made to the control policies described in Chapter 8 in order to reduce the rapid channel-switching operations at the cost of some loss of responsiveness. These are the *moratorium* and *delayed close* policies.

9.3.1 Moratorium policy

A moratorium policy is a policy where a *moratorium* is placed on any change in the bandwidth for a period of time τ after a successful increase/decrease in bandwidth. Such a policy will reduce the number of transitions in bandwidth over a given period of time (or the number of switchings per packet per unit time) compared with a similar policy without a moratorium operating under similar conditions (arrival and service rates, threshold levels, etc.). However, this reduction in the number of switchings will be achieved at a cost to the responsiveness of the system to fast changes in the input traffic rate. From control theory, this seems to be a valid trade-off.

Although not taken further in this book, an interesting point would be to find out if an optimum value for moratorium exists for a given arrival and service rate combination. Table 9.2 shows the effect of moratorium policy (adapted for policy type 1) on various system parameters. It can be seen that the effect of a moratorium is to decrease the number of switchings per packet, and to increase the response time and the cost of the bandwidth used. This indicates that a moratorium policy may be used in situations where the cost of switching is comparatively higher than other costs.

Table 9.2 Effect of moratorium policy on system parameters[a,b]

Measure	$\tau = 0$ ms	$\tau = 100$ ms	$\tau = 1000$ ms
N	5.344	5.414	6.258
Q	4.456	4.521	5.376
RT	334.602	338.612	390.702
SW	0.182	0.157	0.053
BW	2.256	2.247	2.318
C_{BW}	140.87	154.34	159.24

[a] Using policy 1, with $\lambda = 16$ pps, $\mu = 8$ pps, $UL = 4$ BBUs, $H = 10$, $L = 0$ and $SW\_T = 10$ ms
[b] N: number of customers in system; RT: response time; UL: bandwidth upper limit; τ=moratorium time; C_{BW}: cost of bandwidth; BW: mean bandwidth used; SW: number of switchings/packet

9.3.2 Delayed close policy

The delayed close policy is a type of one-sided moratorium policy; that is, it puts a moratorium on the *decrease* of bandwidth of a channel whose service rate is under control; the increase in the bandwidth is not affected by this policy. One advantage of a delayed close policy is that the moratorium delay can be linked to the tariff structures of the underlying communications link.

The ISDN tariff structure is described in Section 2.10 and its effect on channel management is discussed in Chapter 4. According to these, it pays to keep a channel open until the end of the current charging period once it is set up. There are no guidelines as to how a channel connection cost will be calculated when the channel capacity is allowed to vary dynamically during the lifetime of a connection. If the superchannel is constructed by the aggregation of n x B-channels (described in Chapter 6), then each B-channel will be charged individually. It pays then to monitor the connection and charging periods of each B-channel comprising a superchannel for closing the B-channel with the most suitable charging period so as to minimize connection costs.

A simplifying assumption can be made when modelling this case with regard to the charging periods. It may be assumed that, every time an increase in the channel bandwidth is achieved, the charging period is re-synchronized to this time (i.e., the charges accrued so far are added up and a new charging period started); hence all the calculations for the close delay can be calculated from this time. Appendix V shows pseudo-codes implementing delayed close policy where it can be applied to the last remaining bandwidth unit of the superchannel or to any bandwidth decrement operation. This routine will be activated every time the queue length check indicates the low threshold triggering. It will then invoke the function *delayed_close()* to determine whether to close the channel immediately or after a delay.

A further condition set into the delayed close function could be that no further packet arrivals should occur during the active period of delayed close operation. If a packet arrival occurs, then the delayed close would be aborted. This may be imposed so that a channel is not closed if packets continue to arrive after the triggering condition for delayed close. The cost performance of the delayed close policy would be heavily dependent on the channel cost per unit time and on how heavily the switching operation is penalized in terms of cost and time.

9.4 BI-LEVEL V. MULTI-LEVEL CAPACITY CONTROL

As can be seen from Figs 9.1, 9.3, 9.11 and 9.12, both the bi-level and multi-level control policies perform reasonably well within the limits of their control ranges. Multi-level control provides a smoother transition between the lower and higher service rates; that is to say, it operates with a finer granularity. The only peculiarity of the multi-level control appears to be situated around $\rho = 1$ point, i.e. the capacity of a basic bandwidth unit (BBU). Around this range, the mean number of customers in the system, and hence the mean response time, both increase quite steeply. Beyond this point, the number of customers in the system remains quite static while the mean response time falls with an increase in the arrival rate (as N/λ).

Figures 9.1 and 9.3 shows that the bi-level (overload/underload) control system does not suffer from the critical behaviour around the $\rho_0 = 1$ for values of $\rho_1 < 1$. It would, however, suffer from the same problem if $\rho_1 = 1$. It also suffers from the same degradation in performance, as its saturation point is reached at the ρ_1 = maximum value (see Figs 9.1 and 9.3).

In Table 9.1 the multi-level Policy 2 and its bi-level counterpart are compared. When the bi-level policy is operating at the $n \times$ B-type interface, the number of switchings required and the response time are larger than for the multi-level policy. However, bandwidth holding costs are slightly less than for the multi-level policy. If the bi-level policy is operating at the $n \times 64$ kbps type interface, all the additional servers needed for ρ_1 can be obtained in one set-up request, and hence the number of switchings will be reduced.

We can conjecture that the multi-level control policy would perform better in the case where multiple parallel superchannels sharing the common interface are competing for the bandwidth. An example can be given to validate this conjecture. Assume that an interface exists with 6 BBUs and that two distinct and competing superchannels are operating in parallel subject to threshold service capacity control policies. Two cases can be visualized with regard to the number of control levels at which the policies are operated:

- *Case 1* Both superchannels using bi-level control with levels set to 1 BBU and 4 BBUs
- *Case 2* Both superchannels using multi-level control with levels set to 1 BBU (minimum) and 4 BBUs (maximum), with the intermediate 2 and 3 BBUs also available in this case

For both cases, assuming a mean traffic arrival rate per superchannel (SC) to be equivalent to less than 3 BBUs (i.e., 3 BBUs would be satisfactory for performance purposes), in case 1, as one SC grabs 4 BBUs, the other SC will have to wait for the release of the BBUs so that it in turn can grab the 4 BBUs necessitated by the traffic arrival rate. This means that one SC will be blocked while the other grabs the available bandwidth. During the time the second SC is blocked, one BBU will be unusable and hence wasted. In case 2, however, since all the levels between 1 and 4 BBUs are allowed, the two SCs will operate at any bandwidth in this range as the arrival rate dictates.

9.5 PRACTICAL ASPECTS

Both the bi-level and multi-level control doctrines would be overwhelmed if traffic intensities greater than their static control ranges were to occur. This can be prevented by having an *adaptive scheme*, whereby the control range is varied according to the congestion at the queue as well as the bandwidth available at the interface and the contention for it. Two methods can be used for the adaptive mode of operation:

1. *Multiple thresholds* For a two-threshold pair system, the higher threshold of the second pair could be set to a value greater than that of the higher threshold of the first pair while the lower thresholds could be identical. If the number of customers in the queue grew to the higher value of the second threshold, this could trigger a higher $\rho_1^{\prime}$ value for the bi-level control, which would then be kept until the number in the queue was reduced to the lower threshold. At this point the system would revert back to the (lower) ρ_1 value. A major problem with this method is that, when the higher service rate request of this queuing system is blocked as a result of bandwidth contention, its queue length will grow with time, causing it to cross one or more higher threshold pairs. This will then precipitate further blocking since the higher thresholds will need more channel capacity. In the multi-level control, this problem will not occur to the same degree.
2. *Variation of the control metrics* This method necessitates the variation of the values of the ρ_0–ρ_1 pair in the case of a bi-level control doctrine, and the variation of the threshold levels rather than their form in the case of a multi-level control doctrine.

9.6 SUMMARY

Service capacity control doctrines based on the threshold control principle can be classified into different categories according to the form, number of control levels and number of thresholds (or threshold pairs). The form of the threshold policy may be simple or hysteretic. The number of control levels can be uni-level, bi-level or multi-level. The number of thresholds or threshold pairs can be single or multiple. Threshold-based service capacity control, whereby the increase in the service capacity is achieved after a random delay (exponentially distributed with mean rate γ), has recently become of interest as a result of the ISDN multi-channel capabilities and channel-switching delays. Bi-level and multi-level service capacity control form an interesting class of policies for threshold control of queuing systems.

A subtle modification to the multi-level service capacity control doctrine introducing the moratorium policy is shown to reduce the number of switchings per packet per unit time at the cost of reduced responsiveness of the control system. A delayed channel close policy is proposed for utilizing the whole of the charged period in an expensive tariff environment. We conjecture that the multi-level service capacity control is more flexible in an environment where multiple parallel superchannels compete for the available bandwidth at a common interface. Analytic results for their relative performance under such conditions remain to be developed. However, the advantage of the multi-level service capacity control over its bi-level counterpart under certain conditions is demonstrated by way of an example.

CHAPTER

TEN

RATE-BASED BANDWIDTH MANAGEMENT

In the previous chapters queue-based dynamic bandwidth management policies were discussed. In this chapter a second set of dynamic bandwidth management policies, based on the data arrival and communication channel transmission rates, is examined. These have some similarities to the queue-based service rate control policies using thresholds described in the previous chapters. The arrival rate of data input (traffic or load) to the controlled system is calculated by measurements taken at regular intervals, and this is then compared with the transmission rate or some threshold values. The decision to keep, to increment or to decrement the existing aggregate channel capacity is then based on some threshold, timer or tariff function. The exact system state index used in the control mechanism varies with the method adopted. However, the aim of all the management policies is to allocate as much bandwidth for as long as possible to the requests so that, on the one hand, an acceptable performance in terms of delay and throughput is achieved and, on the other, unnecessary channel-switching and holding costs are minimized. Several rate-based bandwidth control methods are considered: namely, the loop control method, the rate weighting method and the rate thresholding method.

10.1 LOOP CONTROL METHOD

This is a model based on the control theory where the output is driven by the input and the current state of the system control variables. The data input arrival rate, current settings of the output channel capacity and control variables are all taken into account in making a decision to increase, decrease or keep the same channel transmission capacity (Altarah and Seret 1991). Figure 10.1 depicts the loop control model. The configuration results in a 'negative feedback' control system. This is a discrete-time model such that the system state is sampled at regular intervals, for example every T seconds, and decisions are made at the end of the sampling period. This produces an output decision mechanism which always lags the input mechanism.

A system where the output tries to copy the input in a delayed but verbatim fashion may oscillate. However, by the use of some suitable modifications to the model, a more stable control mechanism may be developed. These are discussed further below.

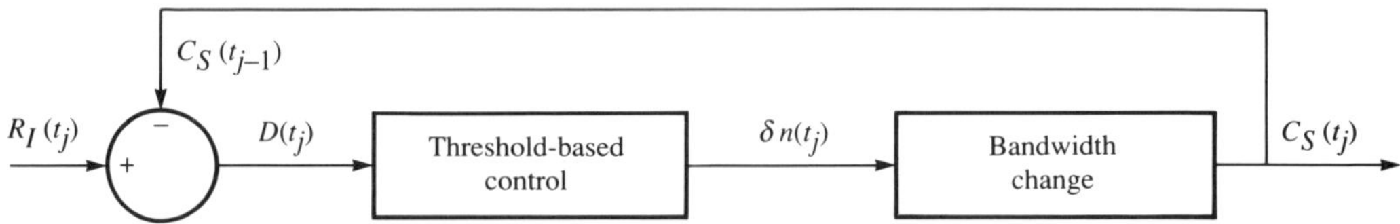

Figure 10.1 The loop control method

The loop control model is based on the comparison and evaluation of the difference between the data input and output rates periodically at times t_i. The data input rate, measured in the interval t_{j-1} to t_j, is given by R_I and is measured in bits per second. The output of the control loop is the channel transmission (i.e. service) capacity in bits per second, given by $C_s(t_j)$. The following additional definitions are made:

- $C_s(t_{j-1})$: service capacity $n(t_{j-1})C$ at time t_{j-1} (in bits per second)
- $n(t_{j-1})$: the existing number of channels at time t_{j-1}
- C : capacity of a single B-channel (56/64 kbps), or basic bandwidth unit (BBU)
- t_j : time at jth sample; for periodic sampling with period T, this is jT seconds from the start of the control period
- $\delta n(t_j)$: The number of B-channels to be set up or cleared down

Therefore, the difference between the input and output data rates is given by:

$$D(t_j) = R_I(t_j) - C_s(t_{j-1}) \qquad (10.1)$$

The threshold-based control module in Fig. 10.1 calculates the next action to be taken regarding the bandwidth change. The action is indicated by the variable $\delta n(t_j)$, and it can have a positive or negative value. A positive value indicates set-up and a negative value indicates removal of δ number of B-channels.

The simple difference method is not a good metric for making the decisions of bandwidth management since it indicates system evolution only within one sampling period, that is T seconds. A better metric to use is to sum this value over the past history of the system evolution. Equation (10.2) gives the value of $S(t_j)$, the sum of differences from the first to the current sampling period (Altarah and Seret 1991; Buhler 1982). It has been shown that this gives a stable system control under linear traffic assumptions (Altarah and Seret 1991):

$$S(t_j) = \sum_{i=1}^{i=j} w_i \, [D(t_j)] \qquad (10.2)$$

where w_i is the weighting factor for the ith difference and $D(t_j)$ is given by (10.1).

10.1.1 Main features of the control system

A general threshold control function is shown in Fig. 10.2(a). The decisions for the establishment of new B-channels to increase the aggregate channel bandwidth and release the existing B-channels to reduce aggregate channel bandwidth can be based on the values of two thresholds: T_e, the *channel establishment threshold*, and T_r, the *channel release threshold*. A more general control policy may include the case where both T_e and T_r, are functions of the system state. This could be used, for example, in making new channel establishments more difficult once a certain number of B-channels are already allocated, and in making removals of existing channels more difficult once a minimum number of channels are left in the aggregate channel. These can be shown as $T_e[n(t_j)]$ and $T_r[n(t_j)]$ in Fig. 10.2(a).

Another input to the control mechanism is the *mode control*, MC, which decides whether single or multiple B-channels (or basic bandwidth units—BBUs) can be added to or removed from the aggregate channel. A control value of S or M can be used to indicate single or multiple channel actions. An added control mechanism that can be used with mode M is v, a control variable indicating the ceiling value for the number of B-channels that can be switched at one time in multiple channel action mode. It can be argued that the multiple action policy is more efficient in dealing with highly bursty traffic types (see Altarah and Seret 1991).

Figure 10.2(a) also shows a control function based on the observed *rate of change* (RoC) in the difference value between the data input rate and the channel transmission rate in bits per second. This is practically the last value of the $D(t_j)$. Its involvement in the decision mechanism can provide responsiveness to the control mechanism. Since each channel establishment takes a finite time (for example, 0.5–3 s in ISDNs), it may be necessary that a moratorium is placed on any further B-channel establishments until the ongoing channel establishment has been completed. This is shown in Fig. 10.2(a) as T_m. For practical purposes, however, a much more simplified threshold control form can be used as indicated in Fig. 10.2(b).

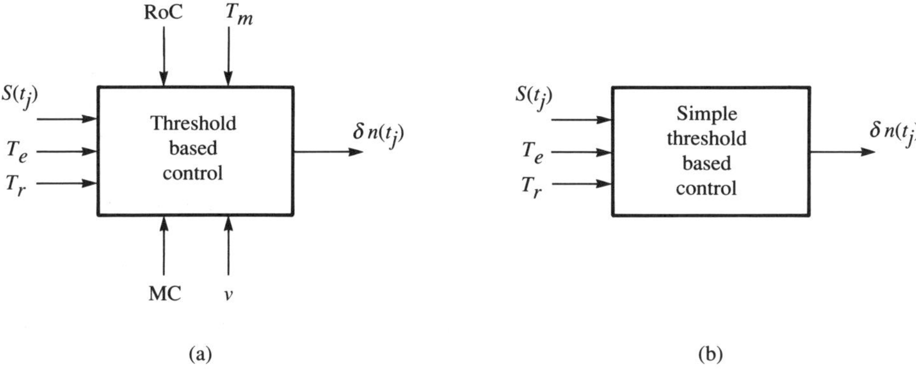

Figure 10.2 Threshold control functions: (a) general; (b) simplified

Figure 10.3 shows a sample variation of the aggregate channel capacity in response to the variation of the input data rate. Example settings could be R_j = 256 kbps, T_e = 64 kbps, T_r = 64 kbps, w_i = 1 and T = 1 s.

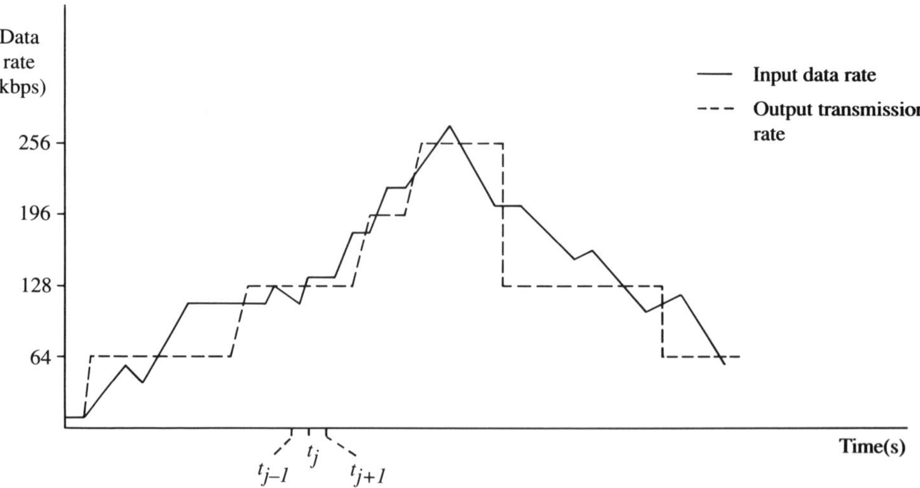

Figure 10.3 Variation of aggregate channel capacity in response to variation of input rate

Figure 10.4 shows a sample variation of $S(t_j)$—the sum of differences from the first to the current sampling period given by (10.2) and (10.1) versus time. The figure also displays a sample variation of the data rate versus time.

10.1.2 A case for compromise

It is apparent that a relationship between the responsiveness of the control system and the cost of operation exists. For example, it is desirable that the system follows the demand from the input traffic changes in either direction (decrease as well as increase) so that queuing delays can be minimized. However, because of delays in the channel set-up and cost of switching a new channel in, it is sensible to delay the switching off of any aggregated channels. This means that the channel holding cost will go up but that it will also reduce the queue size. Another criterion is the system stability. An over-eager threshold setting may cause oscillations to occur when a better policy would keep the number of channels steady. These are discussed further below.

10.1.3 Analysis of major findings

Some of the main findings of tests conducted by Altarah and Seret (1991) using the control loop methodology include the following:

1. Channel establishment and release thresholds do not necessarily have to be equal. Indeed, the channel release threshold is expected to be greater (say, three times) than the establishment threshold.

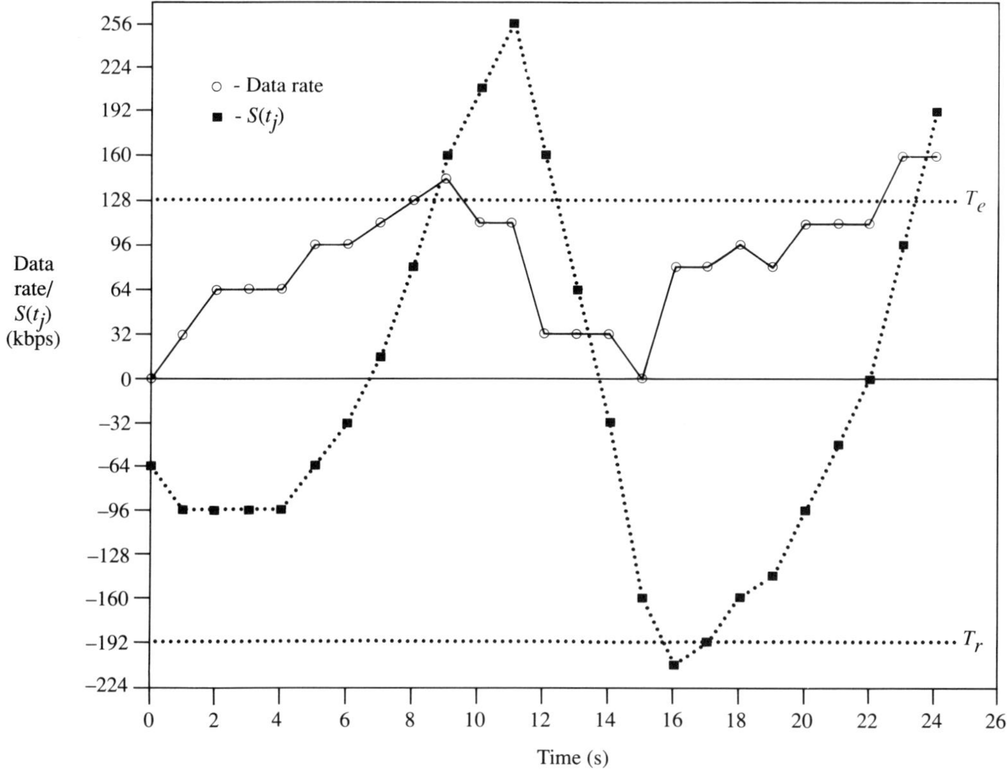

Figure 10.4 Variation of $S(t_j)$ with changes in data rate ($w_i = 1$; T_e = 128 kbps and T_r = 196 kbps)

2. A good policy using the control loop methodology is to have adaptive threshold settings for the channel establishment, T_e, and channel removal, T_r. One policy could be to increase the establishment and release thresholds by 64 and 128 kbps, respectively. The values should be reset to their initial settings every time the system crosses between the establishment and release cycles.
3. The sampling period T must be long enough to allow the system response to settle and short enough to permit responsiveness. Large values of T result in long queuing delays.
4. A careful choice of the difference weighting factor w_i is important in preventing premature channel release. A small value will ensure that the system will reach the release threshold value more slowly. This will prevent the establishment of unnecessary channels later on and will reduce the data queue size, improving the average queuing time.
5. The value of the channel set-up delay has been found to have no effect on the system output behaviour. It did however, increase the mean response time.

Other findings by Altarah and Seret are summarized in Table 10.1.

Table 10.1 Comparison of applications over aggregate channels

Application	Finding
File transfer	Small thresholds reduce the system response time at the cost of increased instability.
Pulse image transfer	Small establishment threshold values improve the system's response time. Small release threshold values reduce cost at increased system instability.
Image transfer	Establishment threshold affects the response time as well as the number of transitions between establishment and release, while the release threshold affects channel holding costs and the number of transitions between establishment and release for a given traffic.

It can be shown that two major sources of problems arise with this type of control algorithm:

1. The value of the arrival rate exactly or very nearly equals the channel transmission (service) rate. In this case from the classical queuing theory, operating with a traffic intensity ρ approximately equal to 1 will create long queues, and when exactly equal to 1 will grow unbounded (Gross and Harris 1985). Therefore, care needs to be taken in determining the next bandwidth value to be assigned to the server.
2. Care may be needed when only one channel exists and the rate of arrival is very much below the 64 kbps mark. In this case, channel removal and establishment may be more costly owing to the increased queuing delays. The use of adaptive thresholds will alleviate this problem in many cases.

10.2 RATE-WEIGHTING METHOD

This method is based on the weighting of the successive traffic input rate measurements which are taken periodically, once every T seconds. The weighted rate of data input is then used as the metric that determines whether a new channel is added or removed or no action is taken. The policy is based on the following equation:

$$R^{(i)} = \alpha R^{(i-1)} + \beta R_s^{(i)} \tag{10.3}$$

where $R^{(i)}$ is the weighted traffic arrival rate at time T_i
$R^{(i-1)}$ is the (previous) weighted traffic arrival rate at time T_{i-1}
$R_s^{(i)}$ is the (latest) sampled traffic arrival rate (bit received in sampling interval/interval) at time T_i
α and β are the weighting coefficients where $\alpha + \beta = 1$; typical values for α could be in the range 0.7–0.9.

Equation (10.3) can alternatively be expressed as follows:

$$R^{(n)} = \beta \sum_{i=1}^{i=n} \alpha^{n-i} R_s^{(i)} \qquad (10.4)$$

where n indicates the nth weighting (at the nth sample). This can also be expressed as:

$$R^{(n)} = \beta \{ \alpha^{n-1} R_s^{(1)} + \alpha^{n-2} R_s^{(2)} + \ldots + \alpha^0 R_s^{(n)} \} . \qquad (10.5)$$

It can be seen from (10.5) that the most prominent factor is the latest (nth) sample. The older samples diminish in importance as they are multiplied by increasingly smaller weights. This process has the effect of smoothing out the effect of the sudden changes in the arrival rate on the decision metric.

10.2.1 Characteristics of the rate-weighting method

Some of the most important findings of a simulation study carried out by using (10.3) (Du and Knight 1991) are discussed in the following sections. The test environment in which the method was considered and the main decision mechanisms are described below.

The test environment A suitable test bed for the above bandwidth management method is shown in Figs. 10.5 and 10.6. Figure 10.5 shows the details of the simulation environment used (Du and Knight 1991). The experiment is necessarily done in a simplex configuration with one packet generator and sink. The packets are generated according to a selected distribution for inter-arrival time, for example negative exponential or hyper-exponential distribution. A channel management module makes the necessary management decisions. It has control over the D and B-channels at the interface with the ISDN. The packets travel over the channels assigned and reach the destination gateway which forwards the packets to a packet sink. At this stage some statistical data is collected for later analysis. Packets reaching the sink are then simply discarded.

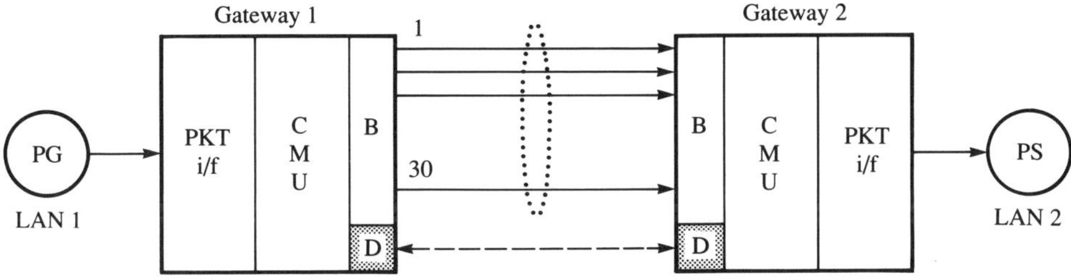

PG: packet generator; PS: packet sink; B/D: user/signalling channel
CMU: channel management unit; PKT i/f: packet interface

Figure 10.5 The simulation environment

Figure 10.6 shows the details of the channel queues. The output section is an n x A/B/1/K type queuing system. The queue size is limited at both the input and output queues. This means that, if there is not enough bandwidth between the two gateways, some of the incoming data packets will be dropped and lost, causing re-transmission.

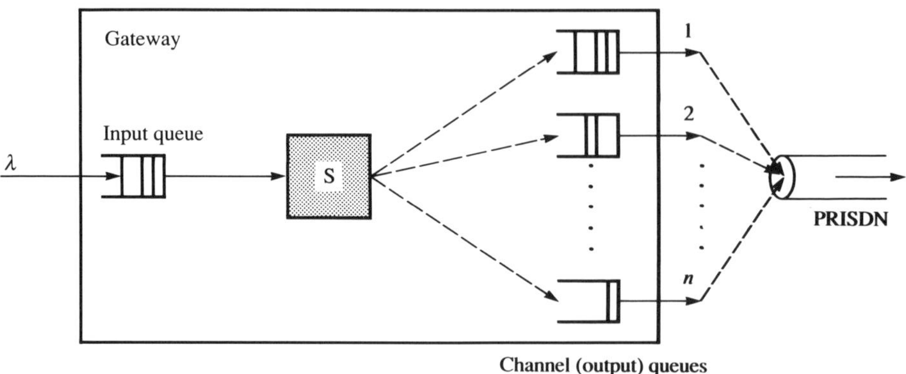

S: packet scheduler; λ: packet arrival rate
PRISDN: primary-rate ISDN

Figure 10.6 The channel queues

Decision mechanisms

New channel establishment policy When a data packet arrives and is addressed to a destination to which no links are already established, a new channel is set up if there are free channels left in the unallocated B-channel pool.

Aggregate channel bandwidth increase This is based on two factors: data input rate measurement by periodical sampling of the incoming traffic, and the threshold-based switching mechanism. At the end of a sampling period, the gateway collects arrival rate information and compares the data arrival and transmission rates. If the difference is above a certain threshold value, then one or more B-channels are set up and aggregated to the existing channel capacity on the next packet arrival epoch.

Threshold settings The threshold setting for channel establishment at time j—that is, the jth sample—$T_e(j)$ is given (in kbps) by:

$$T_e(j) = n(j-1)\,64 + X \qquad (10.6)$$

where $n(j-1)$ is the number of channels existing at time interval $(j-1)$ to (j) and X is the threshold margin over which new channel establishment will be triggered.

Channel removal policy (bandwidth decrease) This is closely linked to the packet scheduling and the tariff-based policies. It can be based on the *channel idle time*. If a channel is not utilized within τ seconds of its last use, then it is deemed appropriate to close that individual channel. This policy of channel removal can work only with an 'uneven' packet scheduling scheme, owing to the above condition; it will not be appropriate for use with, say, random or round robin packet scheduling schemes. This type of scheduling has been mentioned in the previous chapters (see Section 4.6) and was called *sequential* scheduling.

Packet scheduling policy A sequential packet scheduling policy is used in such a way that for a group of, say, n aggregated channels the packets are scheduled in sequence starting from channel 1 to channel n, but with a proviso that channel 1 is first filled to the limit of its holding capacity before any packets are queued, with the same principle at the remaining channels.

Tariff based policy Using this policy, ISDN tariffs are taken into account in closing an existing channel. It is known that ISDN tariffs for circuit-switched data follow the same stepped charging policy as that of the PSTN. For example, in the United Kingdom this is 4.4p per charging period of 18 s. Once a new charging period is registered, the whole of the period charge is payable even though only part of the 18 s is utilized. This means that, once a new charging period is crossed, the channel already set up should not be removed until just before the next charging period. This optimizes the service obtained per unit investment.

Sampling period The sampling interval has a profound effect on the responsiveness of the management policy. A too-short sampling period will not capture the 'long-term' trend and will be over-reactive, causing an increased cost of communications. A too-long sampling period, on the other hand, will average out the peaks in the arrival rate, causing the channel management to be sluggish in its response. This will mean fewer channel establishments than should be adequate for a given traffic. This will result in packet losses arising from queue overflow and will cause retransmissions, increasing the delay in communications.

10.2.2 Multiple channel establishment policy

Multiple channel establishment can be achieved by using (10.6) and (10.7):

$$\delta n(j) = \frac{R^{(i)} - Te(j)}{64} \qquad (10.7)$$

where $\delta n(j)$ is the number of new channels that need to be established. It gives the difference in the input and output rates (plus the threshold margin) and scaled by 64 kbps to give the number of B-channel increments necessary at time j.

10.2.3 Analysis of major findings

The main conclusions on findings of tests conducted using (10.3) and the associated control mechanisms (Du and Knight 1991) may be summarized as follows:

- A threshold control value of X in the range 40–50 kbps is found to be reasonable. A value lower than 25 kbps will cause too quick a response to the changes in the arrival rate, such that, once established, they will seldom be used, increasing the channel holding cost.

- At the maximum traffic input rate of 320 kbps, the input queue size of 100 packets (packet length of 256 bytes) will guarantee that the packet drop rate is less than 1 per cent. However, this is also dependent on other factors.
- The channel closing policy based on the idle time does not work effectively if the departure queue size is too small. It is found that a queue size of five packets is adequate.
- A sampling interval value of between 100 and 400 ms is found to be acceptable. A typical selected value would be 200 ms.
- Tariff-based bandwidth management is a sensible choice since it utilizes the bandwidth already paid for (owing to charging period crossings).
- Optimization of the individual control parameters is difficult because of the interreaction between them.
- The decoupling of channel establishment and release, so that they will have a different set of triggering mechanisms, does not produce the fine control that is needed in such applications.
- The use of the next packet arrival epoch as the triggering mechanism causes an asynchronous operation to that of sampling and adds some sluggishness to the system bandwidth change response, especially at low bit rate arrivals.

Figure 10.7 shows the packet delays and bandwidth control using different management policies in the rate-weighting method.

10.2.4 Possible extensions of the model

It is of course possible to use the rate-weighting method with a superchannel giving $n \times 64$ kbps (i.e. A/B/1—$n \times$ C-model where C is 64 kbps) rather than the simulated model where the resultant queuing system is an $n \times$ A/B/1. However, the release of the channels would also have to be related to the rate and/or the queue threshold values.

10.3 RATE THRESHOLDING METHOD

This is based on the application of the hysteresis technique to the decision mechanism for channel establishment and release in dynamic bandwidth management. The simplest form of hysteresis control functions across a threshold pair,[1] as depicted in Fig. 10.8. This is a two-level (phase) hysteresis control method. We assume that the traffic stream is allocated bandwidth B_1 bits per second to start with. When the data input rate reaches the upper threshold from below, the channel bandwidth is incremented to value B_2 bps ($B_2 > B_1$), and when the data input rate falls to below the lower threshold from above, the channel bandwidth is reduced back to B_1. This simple model can be further extended to include multi-level bandwidth switching. This is discussed further in the next section.

[1] An even simpler form of threshold control is one with a single (point) threshold, where the two thresholds are identical.

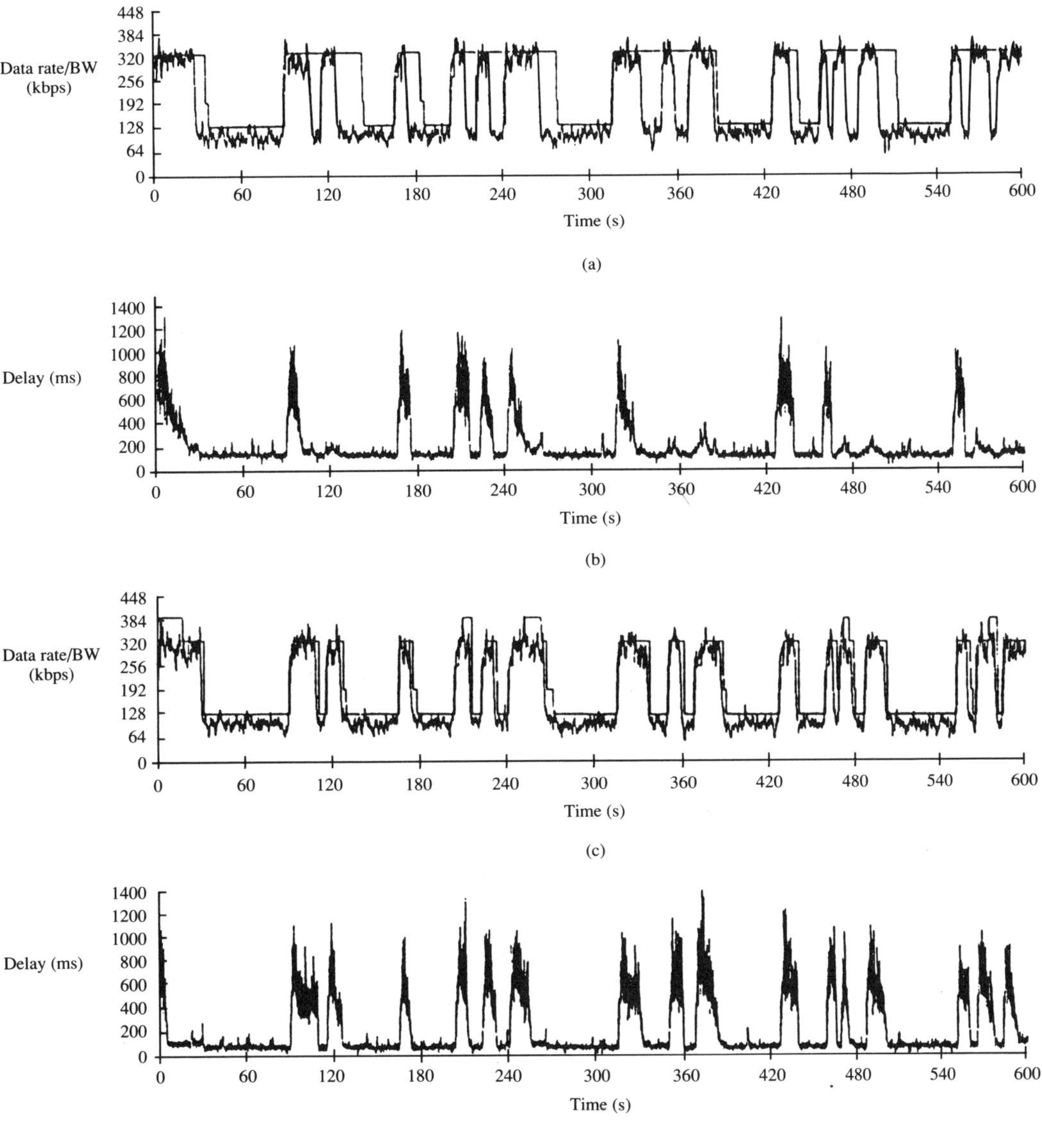

Figure 10.7 Channel management policies based on the rate-weighting method:
(a) LAN traffic arrival rate/ISDN bandwidth versus time (tariff based)
(b) Gateway-imposed delay versus time (tariff based)
(c) LAN traffic arrival rate/ISDN bandwidth versus time (non-tariff based)
(d) Gateway-imposed delay versus time (non-tariff based)
(Reproduced with kind permission of X. Du & G. Knight (1991))

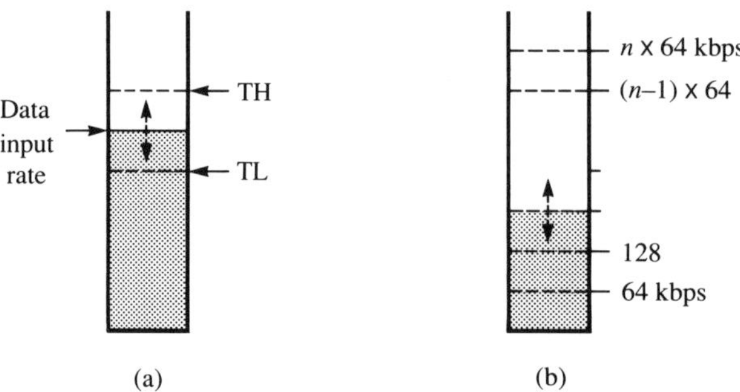

Figure 10.8 Rate hysteresis method: (a) a single threshold pair; (b) multiple thresholds located at natural channel boundaries of n X 64 kbps

10.3.1 Main features

Both the single and hysteresis thresholds can be used for the rate thresholding method of bandwidth control. Similarly, a bi-level or multi-level bandwidth control can be used.

Figure 10.9 shows the rate point thresholding method in action over a two-phased bursty traffic. Figure 10.10 shows the same method over a three-phased load. It has been shown (Harita 1991) that the rate-based policies tend to switch in only the necessary bandwidth that is computed from the data arrival rate. This is similar to the multi-level control used in the queue-based policies described in Chapter 7.

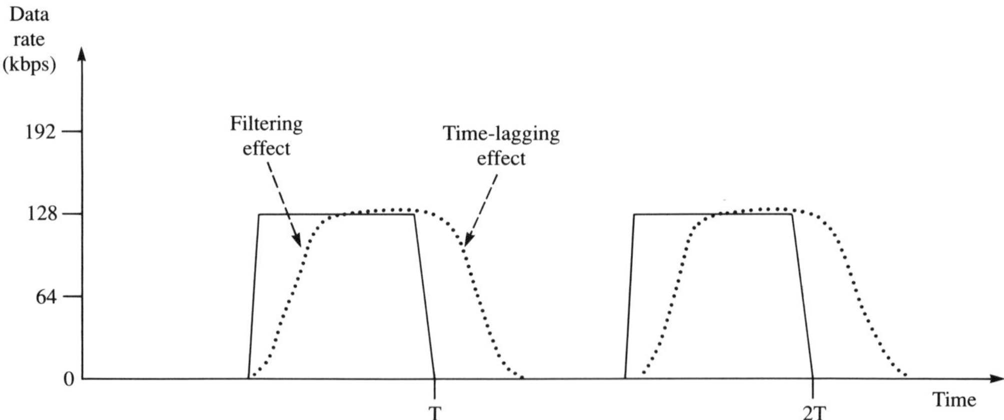

Figure 10.9 Rate point thresholding method with two-phased bursty traffic

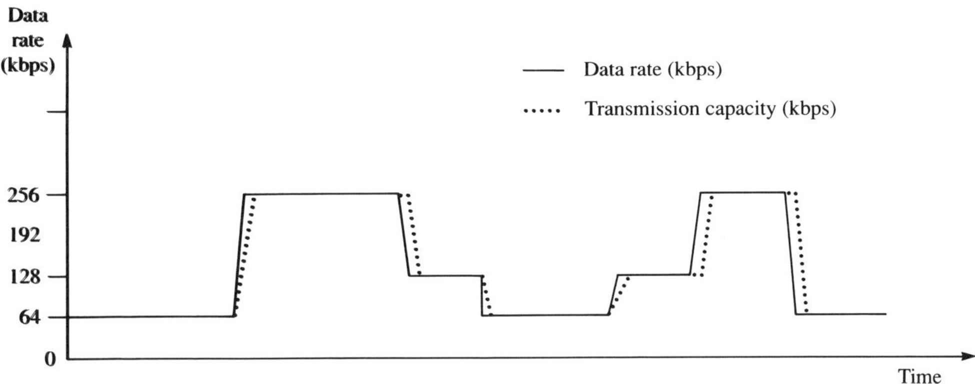

Figure 10.10 Rate point thresholding method with three-phased bursty traffic

10.3.2 Analysis of major findings

Some of the major findings of the analysis carried out by Harita (1991) may be summarized as follows:

1. The bi-level rate-based control across a single threshold pair does not suffer the same disadvantage as its queue-based counterpart; that is, it does not have the danger of constantly adding bandwidth once the threshold has been crossed. (See Section 9.3.1 on moratorium policy.) This is true unless the traffic rate keeps increasing.
2. The reclamation of resources is best done in a delayed fashion so that, once the cost of establishing a new channel is incurred, it is not released prematurely during a temporary lull in the channel operation. This leads to a policy of bandwidth release in single B-channels (or basic bandwidth units).
3. The method is quite easily adaptable to multiple threshold pairs.
4. The rate-based policies are more reactive than the queue-based counterparts. This is due to the explicit measurement of the input rate.
5. The thresholds are naturally set at 56/64 $\times$ n kbps, $n = 1,2,3, \ldots, N$, where N is the maximum permissible number of channels. These threshold boundaries can be used for both the point and hysteresis threshold control policies.
6. Highly reactive policies have been found to be cost-effective when used with short charging periods, whereas those with greater bandwidth retention performed better when used in longer charging periods.

10.4 COMPARISON OF THE RATE CONTROL METHODS

The loop control method is based theoretically on a sound principle: that of closed-loop control theory. It has a negative feedback loop giving it stability over a wide range of input values. However, owing to the fact that the controlled device is a queuing system, it would need to have a special protection mechanism whereby the output is set not exactly to the input value. This is because, in queuing theory, the stochastic nature of the arrival and service processes means that at traffic intensities (i.e. ρ = arrival rate/service rate) close to 1 long queues are formed. The added control would try to set the value of ρ to, say, something between 0.5 and 0.7. This system would then switch in/out the necessary number of channels such that the given traffic intensity is maintained. Indeed, this observation applies to all the other types of rate-based control methods.

The rate-weighting method also uses a form of loop control in that it checks for the difference between the last traffic arrival estimate and the number of channels already existing and then compares this with the threshold value. If the difference is above a threshold value, a decision to establish new channels is made. This process is repeated at every packet arrival epoch. Since in its simplest form it increments the bandwidth with a single BBU (56/64 kbps), the loop is repeated at the next few arrivals of packets. This is in contrast with the loop control method previously described; there, the decision to increment or decrement is done at the end of each sampling interval asynchronously to the packet arrivals.

On the other hand, the rate thresholding method has no loop control as such. Instead, either a simple or hysteresis threshold-based control is applied. No comparison in terms of performance, cost savings or ease of implementation is currently available for these three different methods.

10.5 COMPARISON OF QUEUE AND RATE-BASED DBM POLICIES

The queue-based policies produce a more oscillatory output than the rate-based ones if suitable care is not taken. This is because, even in a steady-state queuing system with Poisson traffic source and exponential service times, the queue length can be described by its mean value as well as its distribution. Furthermore, when operating very close to the traffic intensity value of ρ=1, this distribution is widened. So, even without the changes in the traffic arrival rate, one can observe oscillations, especially if instantaneous queue length is used as the switching metric. A queue-based policy will release established channels if the queue length falls below the release threshold value even if the traffic rate remains unchanged. This has been observed in Chapter 8 and in Harita (1991). The same occurs when the switching threshold is crossed for channel establishment. More and more channels are established (in the multi-level control) as long as the switching threshold is exceeded. A moratorium policy is proposed in Section 9.3.1 above in order to counteract this. A *hybrid* solution is proposed in Harita (1991), in which the decision to change bandwidth is based not only on the queue metric but also on the rate metric. Another metric that can be used in a hybrid environment is the packet loss statistics (Du and Knight 1991; Harita 1991). The queue state information can be used in the bandwidth increase and decrease decisions as a supplement to the instantaneous data arrival rate information. This reduces reactivity to transients in bursty traffic input and prevents early release of bandwidth if there are residues in the queue from the preceding overload session. Performance is improved if queue residue is cleared before bandwidth decrease (Harita 1991).

A rate-based policy, however, will not switch in extra bandwidth arising from small fluctuations in the rate; also, it will retain the switched number of channels as long as the packet arrival rate remains unchanged. Queue-based policies are found to perform better under bursty loads using suitable thresholds (Harita 1991). However, as mentioned above, they suffer from oscillations. Oscillations can be dampened by the use of time averaging, moratorium-based policies and tariff-based implementations.

10.6 OTHER APPLICATIONS OF DYNAMIC BANDWIDTH MANAGEMENT

1. Dynamic bandwidth management can be used in conjunction with access control (see Section 1.6.4) and bandwidth monitoring and enforcement functions to provide a realizable congestion control framework for an ATM overlay[2] running over a public PRISDN. This can be achieved by the use of a simple admission control scheme (see also Sections 1.6.4 and 4.3 above), coupled with a dynamic bandwidth control scheme to cope with the bursty traffic sources (Harita 1991). Within an ATM overlay over PRISDN, applications requiring guaranteed service can be allocated bandwidth according to their peak rates and expedited data transfer at the contention periods. Other applications can have bandwidth allocations based on their mean bit rates. This framework has the advantage that traffic requirements need not be known precisely. Indeed, this has been shown to be the case for a connectionless network relay too (see Section 4.3).
2. Bandwidth sharing between superchannels is an application that has been proposed throughout the previous chapters. A similar proposal is made by Harita (1991) for bandwidth sharing between aggregate channels in an ATM overlay over PRISDN. In data or data and voice environments, a *complete sharing* scheme for bandwidth access control is adequate. In an ATM overlay type environment, Harita has shown that a bandwidth access scheme based on the peak data rates of continuous bit-stream applications (e.g. voice) and on mean data rates of bursty data are practicable. This type of bandwidth sharing is possible since there are some residual capacities in each allocation arising from the course granularity of the 64 kbps basic bandwidth unit. It is also shown that this sharing will equalize traffic performance and provide fairness between end-to-end applications.
3. The residual bandwidth in a packet voice-data multiplexer has also been shown to be amenable to exploitation in an integrated services environment. This is because the packet voice from a number of sources, when multiplexed together, would yield an instantaneous residual bandwidth at their silence and talk-spurts which will be variable. It has been demonstrated by practical experiments that bandwidth management conserved bandwidth at reasonable system performance for data traffic when it is multiplexed with the voice traffic (Harita 1991).

[2] ATM overlay: provision of an ATM-type facility over another network. This is suggested by Tennenhouse and Leslie (1989) in the context of PRISDN. This is provided by using fixed packet sizes (cells), each of 32–120 octets (Harita 1991). Dynamic use of the channels in a PRISDN is shown to be one way of providing this facility in the meantime. Note that the reverse is also being proposed, i.e. the provision of synchronous transfer mode over ATM networks (such as circuit-switched service).

10.7 SUMMARY

Rate-based dynamic bandwidth management is a viable proposition in environments where there is intermittent use of the channel or where the rate of the arriving traffic is changing. It has been found that, with such dynamic bandwidth management, close control of the bandwidth resource is possible, providing improved performance and reducing costs in many cases. It has the added advantage that no previous knowledge of the traffic characteristics is necessary for effective bandwidth control. The control system either keeps the previous system state information algorithmically, or does not need the previous state information to make its decisions. In either case, bandwidth utilization is increased while resource sharing is achieved.

APPENDIX

I

MODELLING AN ISDN INTERFACE

The modelling of communications systems is a necessary and useful process in the application of queuing theoretic and/or simulation techniques to an analysis of the performance of a design or system. This appendix describes the modelling of a well-established port interfacing technique found in many modern communications equipment. This type of technique is assumed to exist in the Ethernet–ISDN relay described in Chapter 5 and the simulation model used in Chapter 8.

An ISDN interface is a collection of hardware and software elements providing a communications function between the system interfaced and the ISDN.

I.1 HARDWARE INTERFACE

For simplicity, it can be assumed that the processing node or system is composed of a single main processor on a *host processor board (HPB)* and the communications ports are formed by the attachment of single or multiple communications *port interface boards (PIBs)* to the system through a *bus* architecture. Such system architecture is shown in Fig. I.1. The host processor and the interface port processor can communicate with each other using a program that resides in the host memory, called the *port driver (PD)*. The PD separates the hardware-specific features from the host by providing a standard software interface to the host. A set of primitives aid the operation of the PD and the host as well as the *port interface processor (PIP or PP)*. A *port command (PC)* capability gives the host processor the ability to instruct and communicate with the PP.

At the software initialization, a sequence of port commands is issued; these prepare the port interface board for the Transmit (Tx) and Receive (Rx) services. The host–port processor communications carried out is through *dual ported memory*, or what is sometimes called the *global (shared) memory* locations. A typical host–port processor communications data structure set-up is shown in Fig. I.2. The communication between the host and the PP is achieved by the use of:

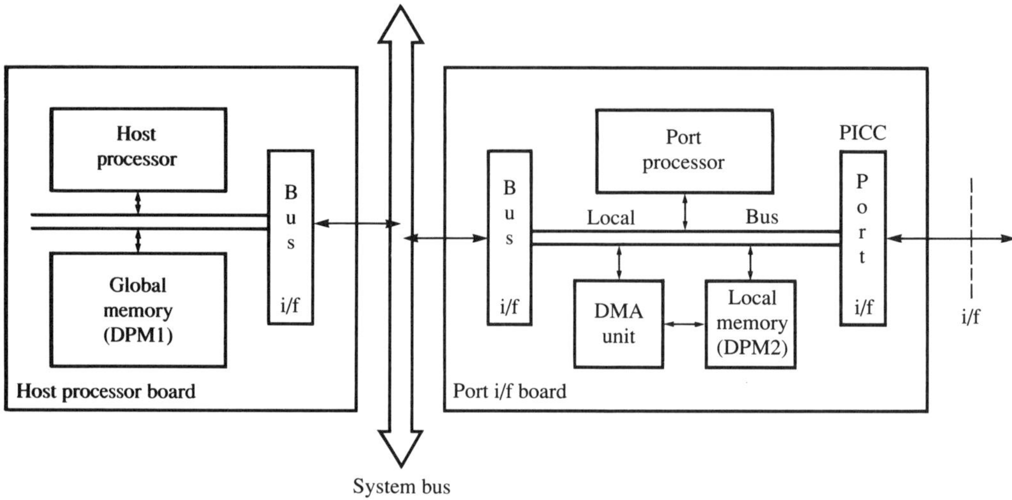

DPM 1/2: dual ported memory 1/2; if: interface
PICC: port interface communications and control unit

Figure I.1 A simple port communication system architecture

- Port control block shared data structures (PCBs)
- Port control registers (PCRs)

A similar set-up is available for communications between the PP and the *port interface communications and control unit (PICC)*. This unit has a set of communications VLSI chips which are capable of handling a multitude of channels as well as some on-board memory, *direct memory access (DMA)* and *synchronous data formatting (SDF)* capabilities (AT&T 1989). Again, the communications is achieved by using some control block and shared memory locations. Transmit and receive descriptor rings are usually used to form a linked list of transmit and receive buffers. The control block may have sub-control blocks for chip attention, channel configuration and interrupt queue pointers.

Chip sets are now available for ISDN interfacing which are capable of operating multiple channels in duplex fashion as well as providing arbitrary and dynamic time slot-to-channel assignments (AT&T 1989). Hence they can provide superchannels, data link layer formatting (e.g. HDLC), transparent operation mode, and variable bit rate control on channels. Two different interface configurations can be defined:

1. Single port processor interface board architecture
2. Multiple port processor interface boards architecture

In the first configuration a single port interface board is available and its control is simpler than the second case, where multiple PIBs are available. The second configuration results if the B and D-channel processings are handled by separate dedicated processors and their respective boards. Also, multiple B-channel port processors may be employed for reasons of

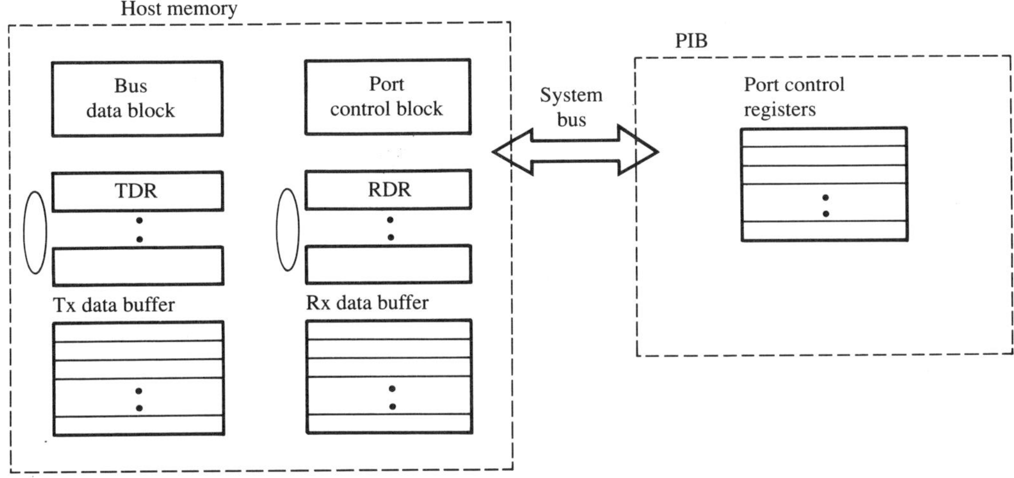

TDR: transmit descriptor ring; RDR: receive descriptor ring
PIB: port interface board; Tx/Rx: transmit/receive

Figure I.2 A typical host-port processor communications data structure set-up

speed, and this could result in multiple interface boards (e.g. ESPRIT project PROOF's Ethernet–ISDN relay architecture—see Knight 1989). The second configuration necessitates complex board management for access to the system bus and the shared memory. Hence for simplicity, in the rest of this discussion only the first configuration will be dealt with.

I.1.2 Main assumptions

The following assumptions are made regarding this interface:

1. The PP is fast enough to appear *transparent* to the following operations:
   (a) *Transmission (Tx)* Reading from Tx ring on DPM1 and placing it in the right channel descriptor ring/buffer on DPM2
   (b) *Reception (Rx)* Reading from Rx rings of each channel, which is done on interrupt pulse; service assumed to be instantaneous
2. Each DLL (HDLC) frame carries one internet protocol (IP) packet.[1]

I.1.3 Main duties of the port processor

The following are assumed to be the main duties undertaken by the port processor:

1. On *interrupt* from port driver: **read** from DPM1 to DPM2 the packet to be transmitted:
   (a) Access DPM1 over system bus and copy Tx packet into DPM2 buffer.
   (b) Fill in the Tx descriptor ring for that channel.

[1] The maximum frame length for HDLC on the SPYDER chip (AT&T 1989) is 4–8191 bytes for the Rx frame.

2. On *interrupt* from PICC: **read** from DPM2 to DPM1:
   (a) Access DPM2 and copy Rx packet to DPM1 over system bus.
   (b) Fill in the Rx descriptor ring for that channel.
3. Process D-channel signalling (i.e., service Transmit and Receive signalling queues).
4. Housekeeping jobs.

I.2 SOFTWARE INTERFACE

The software interface is at two levels, the *protocol* and the *process* level. The protocol level interface is mainly at the data link layer (DLL) in the form of framing, sequencing and error recovery. VLSI chips operating at the DLL (e.g. HDLC) are available. These integrated circuits present an interface at the primitives' level to the user or processes.

The process-level interface can operate on interrupt or polling principles. The PP has several processes running concurrently; their number may vary according to the design and operational conditions (e.g. spawning or killing new or existing processes to deal with certain aspects of interface operation). A software structure, where a central control process forms the basic interface to the processes run by the host processor, can be assumed together with B and D-channel processes and a low-level control interface process (LLCIP) for the chip-level interfacing (see Fig. 5.4).

The placement of various channel management functional units is discussed in Chapter 5. According to Fig. 5.4, the main channel management routine (CMR) and the associated channel status (CST) and routing tables are placed in a process that is run on the *host processor (HP)*. The CMR and the CST could alternatively be placed in a process run by the *port processor (PP)*. This would reduce the number of transactions across the system bus and probably speed up the overall system. It has the disadvantage of loading the PP with the implementation of the channel management routine. This could result in overloading when all the channels are active. In this book the CMR is assumed to run on the host processor.

I.3 SIMULATION MODEL

The granularity at which the simulation of any system is carried out depends upon the performance measures to be used as well as the detail required. Often, major simplifying assumptions are made. In our case, with reference to Figs 5.4, 5.9, I.1 and I.2 these are as follows:

1. The packets arriving from the Ethernet can be simulated by a packet generator (PGEN) which can be set to generate packets according to specific distributions (e.g. Poisson). Packets arriving from the ISDN side of the interface and going to the Ethernet can be simulated by a packet sink (PSINK). Both of these features are incorporated within a USER module in the simulator (see Fig. I.3).
2. The host processor runs two main processes: the internet protocol relaying process (IDRP) and the ISDN routing process (IRP). The ISDN driver process (IDP) and the address resolution (protocol) process (ARP) shown in Fig. 5.9 are assumed to occupy very little processing time and hence to be transparent. The resultant system is simulated

in the relay module (RLM). The host processor is shared between the IPRP and IRP in the following manner. Each process has an event queue and is deemed runnable if there is at least one event in its queue. A process is run to completion by reading the next event from its event queue. At completion, the control is passed to the other process if it is marked runnable; otherwise the same process schedules another event. This is *round robin scheduling* with processes run to completion. This is very similar to the real-time executive CMOS (Wilford, 1984a, b).

3. The IRP deals with two types of events: packet data and internal call control (and channel management) messages pertinent to the D-channel signalling (see Figs 5.4, 5.5 and 5.10). These messages are given priority over the data packets by queuing using the last in–first out (LIFO) principle.
4. The processing times incurred by the IPRP and the IRP (see Fig. 5.9) are fixed constants. This may not be true in reality, as the CMR that runs as a function within the IRP (see Fig. 5.4) may take different times to make the necessary table searches of CST and decisions. These timing factors are not simulated in detail.
5. The port processor on the port interface board runs the BCP, DCP and the LLCIP (see Fig. 5.4). These are modelled simply within the ISDN interface module (IIM) in the simulator (see Fig. I.3). The PP is assumed to be very fast, while the BCP, DCP and LLCIP are assumed to execute in negligible time. Hence all of these processes are assumed to run transparently, so that process-swapping and time consumed in dealing with each input/output request is a negligible cause of processor event queuing. Indeed, these assumptions make the simulation model more akin to a multiple-board interface described earlier where two separate processors run in parallel: one for the BCP and one for the DCP.
6. The transactions across the system or local buses shown in Fig. I.1 are assumed to suffer no contention delays and to take negligible time.

The above assumptions allow us to concentrate on the behaviour of the bandwidth management policies rather than the performance of interface hardware and software configurations. The effective simulation model for each node is shown in Fig. I.3. The full simulation model includes two such nodes back to back with a simulated ISDN interconnecting them. All the experiments described in this book are conducted with only one node transmitting and the other receiving (simplex operation).

I.3.1 Simulation timings

Both the IPRP and the IRP processes are assumed to add a fixed service time to each event (packet or message) processed. This can be user-selected, but their values are set to 10^{-9} s. A fixed packet processing time is assumed to be added to every data packet transmitted and received within the PIB unit which is simulated by the IIM module. The received data packets incur an additional interrupt time. Both the service and interrupt times are set to 10^{-9} s. The transmission delays for the D-channel signalling messages are shown in Appendix III, Table III.1. In addition to these, a fixed LAP-D (layer 2) processing delay of 10^{-3} s is added to every D-channel message except the change bandwidth (CH_BW) message described below.

In the case of superchannel experiments where the bandwidth of an existing channel is increased or decreased, a fixed or random switching time can be selected at the run time. The

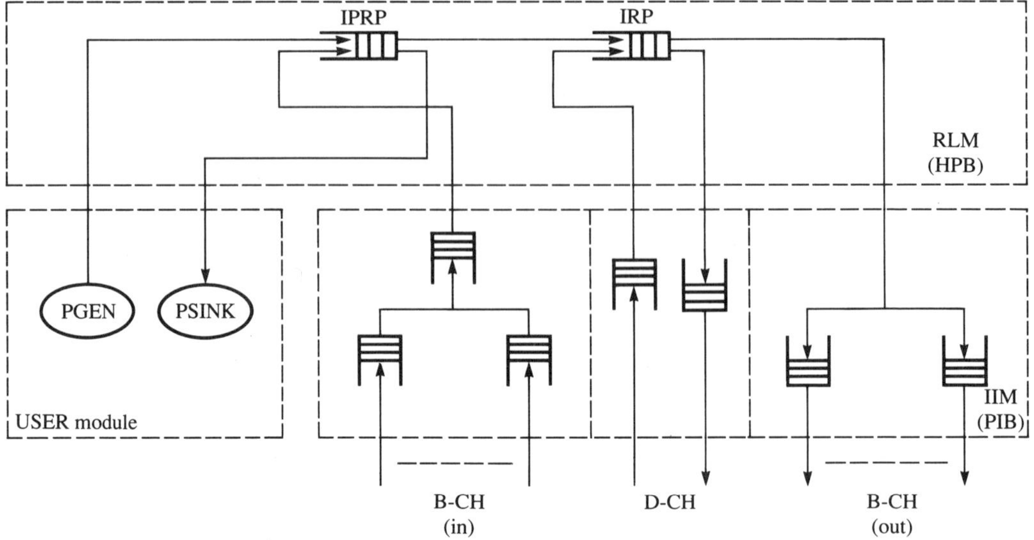

IPRP: internet protocol relay process; IRP: ISDN router process; HPB: hose processor board
PIB: port interface board; RLM: relay module; IIM: ISDN interface module
PGEN: packet generator; PSINK: packet sink; B-CH/D-CH: B/D-channel

Figure I.3 Ethernet-ISDN relay simulation model

random switching time can be generated within the simulator ISDN interface module according to any distribution; however, this has been set to exponential distribution in this book. The bandwidth change of an existing channel between two nodes is simulated by using a special message CH_BW which can affect bandwidth changes in units which are multiples of 64 kbps. As it also incorporates random or fixed switching times, a further LAP-D delay is not added.

I.3.2 Queuing network approximation

The full queuing model of the simulated system, including threshold control of the service rate and incorporating the various processes and their interactions, is difficult to develop and solve in the light of the timing assumptions made in the previous sections. A simplification of the simulated model can be achieved in order to obtain a static queuing model of the system by the following assumptions:

- Simplex operation mode
- Poisson arrivals
- Fixed processing times for the IPR and IR processes or relay module
- Single fixed-capacity channel
- Infinite process queue size
- Error-free transmission channel
- No propagation delay

The first assumption removes the PSINK and B-CH (in) routes of the IPRP of the relay module (see Fig. III.3). The second and third assumptions lead to an M/D/1 model for the relay module. The fourth assumption removes the D-CH loop from the IRP of the relay module, and specifies a single B-CH (out). The last three assumptions simplify the model. Given the third assumption, the resulting *tandem* queuing network is non-Jacksonian (Gelenbe and Mitrani 1980; Gelenbe and Pujolle 1987; Gross and Harris 1985). Furthermore, since not all FIFO stations have exponential service time distributions, the BCMP theorem (Baskett *et al.* 1975; Gelenbe and Mitrani 1980; Gelenbe and Pujolle 1987; King 1990) cannot be used. Since a closed-form solution cannot be obtained, the *decomposition method*, based on the diffusion approximation and developed by Gelenbe and Pujolle (1976, 1987), can suitably be used. The resulting queuing network is shown in Fig. I.4 and the equations applicable for using the decomposition method are shown in equations (I.1) and (I.2).

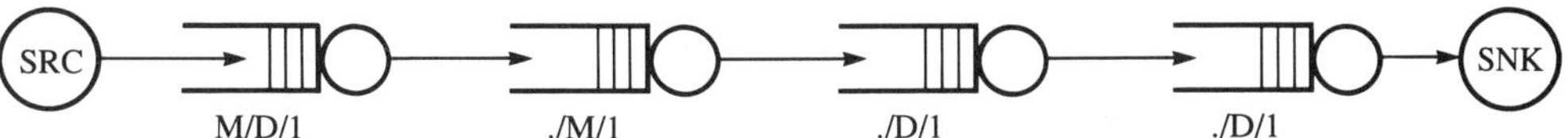

SRC: packet source; SNK: packet sink

Figure I.4 Queuing network approximation

From Gelenbe and Pujolle (1987),

$$Ka_j = \frac{1}{\lambda_j} \sum_{i=0}^{n} \{(C_i - 1)p_{ij} + 1\} \lambda_i p_{ij} \qquad \text{(I.1)}$$

$$C_j = -1 + \rho_j^2 (Ks_j + 1) + (1 - \rho_j)(2\rho_j + 1 + Ka_j) \qquad \text{(I.2)}$$

where Ka and Ks are the squares of the variation coefficients (SVCs)[2] of the arrival and service distributions respectively, C_i is the SVC of the intervals between two departures from station i, p_{ij} is the probability of a packet going from station i to station j, and ρ_j is the traffic intensity at node j. These equations can be used recursively to obtain a solution. The approximate queuing model may also be used to validate the simulation results.

I.4 SUMMARY

The problem of ISDN interface modelling has been described. A plausible model was developed and its basic operation presented. In addition, a simulation model was developed after making assumptions about the nodal hardware and the software. This model is used in the simulation experiments described in Chapter 8. An approximate queuing model for the access node is then obtained by making further simplifying assumptions. This model can be solved analytically by the use of a decomposition method developed by Gelenbe and Pujolle (1987).

[2] SVC = variance/(mean)^2

APPENDIX

II

ISDN SIMULATOR VALIDATION

This section describes the models used for the ISDN simulator (ISIM) validation. The simulator is validated using several models of increased complexity. First, it is validated against an M/M/1 queuing system. Secondly, it is validated against Gebhard's model (1967) which describes the M/M/1 queuing system operating under the bi-level hysteresis service rate control (SRC) doctrine. Third, it is validated against the M/M/c queuing system operating under the multi-level hysteresis service rate control doctrine studied by Moder and Phillips (1962). Although this model is different from our M/M/1 queuing system with variable service capacity operating under a multi-level hysteresis control doctrine, the results obtained would be close enough to show that the simulator is well behaved within the range of inputs used. The following sections provide a comparison of the theoretical and simulation results for the above models.

II.1 THE M/M/1 QUEUING MODEL

This is the simplest queuing model whose performance parameters are given by the classical results (Kobayashi 1978):

$$N = \frac{\rho}{(1-\rho)} \tag{II.1}$$

where $\rho = \dfrac{\lambda}{\mu}$ is the traffic intensity;

$$Q = \frac{\rho^2}{(1-\rho)} \tag{II.2}$$

$$RT = \frac{N}{\lambda} \tag{II.3}$$

where N and Q are the mean numbers of customers in the system and the mean queue length, respectively, λ and μ are the mean packet arrival and service rates (in packets per second), respectively, and RT is the mean response (sojourn) time.

Figure II.1 shows the results of the plot for (II.1) for N versus λ and the corresponding result obtained by using the ISDN simulator. Table II.1 compares values of N for an M/M/1 queuing system obtained by analytical and simulation methods.

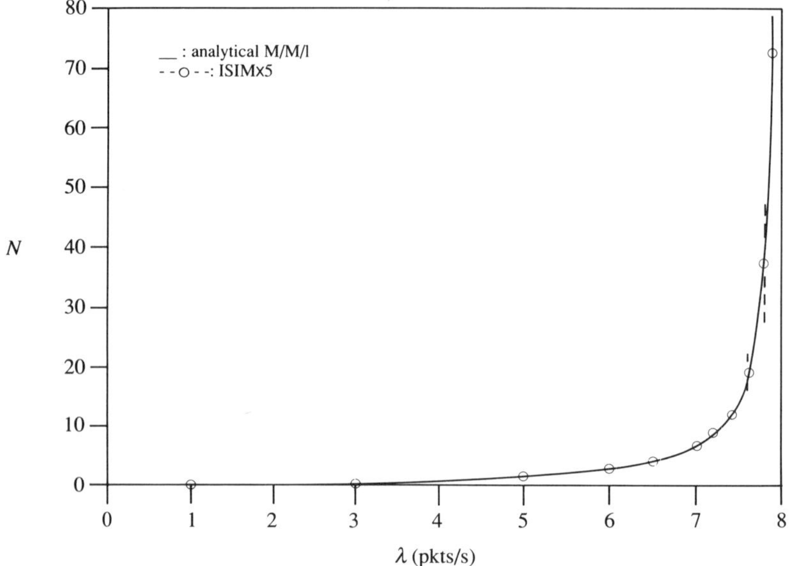

Figure II.1 Mean number of customers in system: N v. λ and the ISIM model (μ = 8 pps)

Table II.1 Values of N obtained for selected values of λ using analytical and simulation methods (μ = 8 pps)

λ[a]	1.0	3.0	5.0	6.0	7.0	7.4	7.6	7.8
Q(analytic)	0.143	0.600	1.667	3.000	7.000	12.333	19.000	39.000
Q(ISIM)	0.140	0.598	1.668	2.942	6.892	12.412	19.464	37.544
Difference	0.003	0.002	–0.001	0.058	0.108	–0.079	–0.464	1.456
Half-width	0.0000	0.0056	0.0345	0.0629	0.3390	1.5297	3.0557	9.7764
Sample var.	0.0000	0.00002	0.00007	0.0025	0.0746	1.5176	6.0559	61.9892
Accuracy	0.0000	0.9286	2.0657	2.1396	4.9232	12.3244	15.6993	26.0399

[a] ISIM: ISDN simulator; N: number of customers in the system; Var.: variance.

All the simulation results are obtained using a sample size of 100 000 packet transmissions and a mean packet size of 1000 bytes for each of the five replications collected for each point on the graph. Furthermore, the timings for processes are set as described in Appendix I. Given that Y_i are the means of individual runs and R is the number of replications, we get the following results (MacDougall 1987):

$$\bar{Y} = \sum_{1}^{R} \frac{Y_i}{R} \tag{II.4}$$

where $\bar{Y}$: sample mean

$$s^2 = \sum_{1}^{R} \frac{(Y_i - \bar{Y})^2}{R-1} \tag{II.5}$$

where s^2 : sample variance

$$H = t_{\alpha/2;\, R-1} \times \left\{ \frac{s}{R^{1/2}} \right\} \tag{II.6}$$

where H : half–width and $(1-\alpha)$: confidence level

$$CI = \bar{Y} \pm H \tag{II.7}$$

where CI : confidence interval

$$RCIH = \frac{H}{\bar{Y}} \times 100\ \% \tag{II.8}$$

Typically, simulation results are required at an accuracy of 10 per cent at the 95 per cent confidence level (MacDougall 1987). Accuracy, or the relative confidence interval half-width (RCIH), gives a measure of accuracy of the estimated mean value. For example, for $\lambda = 5.0$ we can say that we are 95 per cent confident that the confidence interval contains the distribution mean μ, and that we are equally confident that $\bar{Y}$ is within 2.065 per cent of μ (see Table II.1). All the calculations in Table II.1 are made at the 95 per cent confidence level, while the accuracy values are given as percentages and vary with λ owing to the fixed sample and replication sizes.

II.2 M/M/1 UNDER BI-LEVEL HYSTERETIC SERVICE RATE CONTROL

Gebhard's study (1967) provides closed-form expressions for N and Q for an M/M/1 queuing system operating under a bi-level hysteresis service rate control doctrine. A simulation model close to the analytical queuing model can be constructed using the ISIM simulation package. Figure II.2 shows the variation of the queue length, Q, with threshold high (TH_H), where threshold low (TH_L) is set to half the value of TH_H, in both the analytic and simulation models. Table II.2 compares the Gebhard model results with the ISIM results. The simulator results have been obtained by imposing a hysteresis threshold control on the channel queue in the II (ISDN interface) module (see Appendix I). The packet sizes used are exponentially distributed with mean 40 bytes. Packet arrival distribution was set to be exponential at a rate of $\lambda = 800$ packets per second. Under these conditions, the service capacities were set to be 2 and 8 basic bandwidth units (BBUs), where each BBU corresponds to a B-channel capacity (64 kbps). Under these conditions, μ_0 = 400 pps and μ_1 = 1600 pps. Table II.2 shows the results for various TH_H settings where the TH_L values have been set to half those of TH_H throughout. The simulation results are for 100 000 packet transmissions, and the number of replications, R, is set to 10.

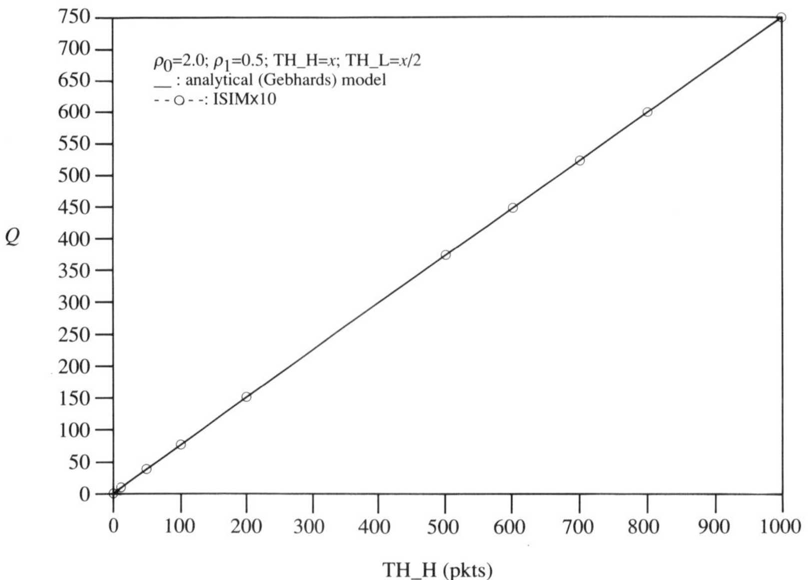

Figure II.2 Mean queue length, Q v. TH_H, for the analytic and simulation models (ρ_0 = 2.0, ρ_1 = 0.5 and TH_L=0).

From Table II.2, it can be seen that the values obtained by Gebhard's formula and those obtained by the ISIM are very close at high values of TH_H. Indeed, the difference between the two sets of values is less than 0.4 per cent of the analytically calculated Q (Gebhard) value[1] for TH_H values equal to or greater than 100. This difference increases, however, to 8.2 per cent for TH_H=10 and to 57.2 per cent for TH_H=2. The difference is not really an 'error' as such. It is attributed to the basic operational differences between the two models, which become prominent at lower values of TH_H. Gebhard's model assumes that the service capacity of the queuing system is increased/decreased instantaneously. In the ISIM model, the increase/decrease in the service capacity becomes effective immediately after the completion of service for the customer in service. As the threshold is lowered, more and more switchings per unit of time take place, hence aggregating the difference observed. Furthermore, as there are now fewer packets in the system, any delay in switching the new service capacity becomes more prominent.[2]

[1] % error is given by (100 x |difference |/ analytical value) and difference = analytical value – (mean) simulation value.

[2] Note that the transmission time for a 40 byte packet at 2 BBUs (128 kbps) is 2.5 ms and at 8 BBUs (512 kbps) is 0.625 ms.

Table II.2 Values of selected parameters for Gebhard's model and the ISIM model

TH_H[a]	2	10	100	500	600	800	1000
Q (Gebh)	1.60	7.0365	74.50	374.50	449.50	599.50	749.50
Q (ISIM)	2.5155	7.6113	74.797	374.65	449.379	599.936	750.06
H (Q)	0.0188	0.023	0.058	0.3978	0.4158	0.5885	0.617
Difference	–0.9155	–0.5748	–0.297	–0.15	0.121	–0.436	–0.56
% error	57.218	8.168	0.398	0.040	0.026	0.0727	0.074
N (Gebh)	2.5091	8.0345	75.50	375.5	450.5	600.50	750.50
N (ISIM)	3.421	8.609	75.797	375.648	450.379	600.934	751.06
RT (Gebh)	3.10	10.0	94.40	469.4	563.1	750.6	938.1
RT (ISIM)	4.2747	10.7555	94.7047	469.355	562.735	750.846	938.454

[a] TH_H: threshold high; TH_L: threshold low (=TH_H/2); N: mean number of customers in the system; Q: mean number of customers in the queue; RT: mean system response time; ($H(Q)$): half-width for Q value.

It has further been found that the small increase in the accuracy of the measured values that would result from increasing the sample size by 300 per cent to 400 000 packet transmissions, or the number of replications by 50 per cent to 15 replications, could not be justified because of the much higher simulation time costs.

II.3 MULTI-LEVEL HYSTERETIC SRC WITH AND WITHOUT SWITCHING DELAYS

An M/M/c queuing system operating under the multi-level hysteretic service rate control doctrine without capacity switching delays has been studied by Moder and Phillips (1962). This model, described in Appendix IV in more detail, bears close resemblance to policy 2 (P2) described in Section 8.5 above. However, the following differences are observed:

1. The Moder–Phillips (MP) model is for an M/M/c queuing system, while the ISIM P2 models an M/M/1 queuing system with variable capacity.
2. The MP model uses Q, the number of customers in the queue, while ISIM P2 uses N, the number of customers in the system, for threshold control of service capacity switchings (increase/decrease).
3. Service capacity increase in the MP model is affected instantaneously (as long as a free server exists) when the high threshold is reached. The ISIM P2 model can increase capacity only after the completion of current packet servicing (and any channel set-up delays).

4. Service capacity decrease in the MP model is different from that in the ISIM P2 model. In the MP model, extra servers are released as soon as each service is completed when the queue length reaches the lower threshold. In the ISIM P2 model, only one server at a time is released at the end of current packet service completion when the number of packets in the system reaches the lower threshold. Each further release necessitates the arrival and departure of another packet and the lower threshold triggering.
5. The MP model assumes the capacity switching time to be zero. The ISIM P2 model can have finite switching times (fixed or random). Indeed, in the experiments described in this section the channel switching time is assumed to be exponentially distributed with mean value of 10 ms.

From the above, we could expect some variations in the performances. Nevertheless, they are found to be close enough to achieve validation over the range of settings. Figure II.3 shows the variation of N with respect to λ for both the MP and ISIM P2 models, while Table II.3 compares the two models. In both cases, the maximum number of servers available (S) is set to 3, while the minimum number of servers (σ) allowed is set to 1. Also, the higher and lower thresholds are set to 100 and 0, respectively (i.e., TH_H=100, TH_L=0). In Fig. II.3 the 95 per cent confidence intervals are shown, while in Table II.3 the accuracy values are given as percentages.

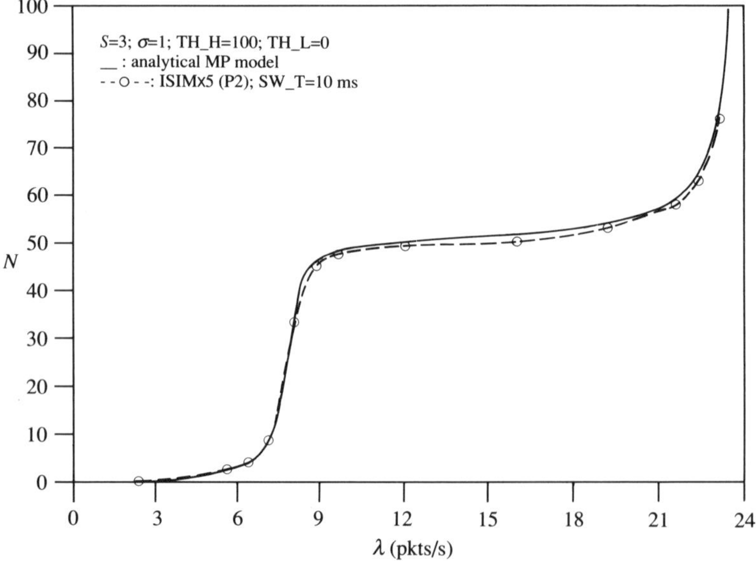

Figure II.3 Mean number of customers in system: N v. λ and the ISIM model (μ = 8 pps)

Table II.3 Values of N obtained for selected values of λ using analytical and simulation methods (μ = 8 pps)

λ	Analytical	ISIM	Difference	Half-width	Sample var.	Accuracy
2.4	0.429	0.43	–0.001	0.0	0.0	0.0
5.6	2.333	2.339	–0.006	0.186	0.000	0.796
6.4	3.999	3.993	0.006	0.041	0.003	1.02
7.2	8.988	8.812	0.176	0.218	0.093	2.472
8.0	–[a]	33.361	–	1.049	2.147	3.143
8.8	46.163	45.767	0.396	0.259	0.131	0.565
9.6	48.395	47.641	0.754	0.377	0.277	0.790
12.0	50.307	49.551	0.756	0.240	0.113	0.485
16.0	52.008	49.830	2.178	0.563	0.619	1.131
19.2	54.267	52.949	1.318	0.361	0.255	0.683
21.6	59.399	57.691	1.708	0.557	0.605	0.965
22.4	64.436	62.462	1.974	1.515	4.485	2.462
23.2	79.470	75.939	3.531	2.644	13.648	3.481

a Note that the (corrected) MP model equation for N (see Section IV.2) does not give the value of N at ρ=1.0.

APPENDIX

III

D-CHANNEL SIGNALLING FOR CIRCUIT SWITCHED CALL CONTROL

This section describes the circuit-switched call control procedures and the expected message transmission delays of the D-channel signalling protocol.

III.1 CIRCUIT-SWITCHED CALL CONTROL IN ISDN AND ISIM

Figure III.1 illustrates an example of call control procedures at layer 3 for a simple circuit-switched call. In the Ethernet–ISDN relay and ISDN simulation (ISIM) used in this book, these procedures have been further simplified for ease of implementation. Hence the following Q.931 (D-channel layer 3) messages are used for simple circuit-switched connection set-up and removal operations in ISIM: SETUP, CONNect, CONNect ACKnowledge, DISConnect, RELease, and RELease COMplete. Note that SETUP ACKnowledge, Information, Call Proceeding and Alerting Messages are not implemented. Indeed, as shown in Fig. III.1, SETUP ACKnowledge and Information messages are optional depending on implementation. Note also that the disconnect phase can be initiated by either party. A further difference between the ISDN and ISIM call control procedures is the end-to-end nature of CONNect ACKnowledge and RELease messages and the local interface use of RELease COMplete to flush buffers in ISIM.

III.2 MESSAGE TRANSMISSION DELAYS

The number of bytes in each message quoted in Fig. III.1 is determined by the size of the *mandatory* information elements and the number and size of the *optional* information elements used. The minimum and maximum allowed values of these messages as proposed in the ECMA-106 standard (ECMA 1985b) are shown in Table III.1. The values of message sizes and their transmission times at 64 kbps (D-channel capacity of PRISDN access) assumed here for the simulation model are also given. Table III.2 shows the corresponding values for the Q.931 settings as defined in the Blue Book (CCITT 1988a).

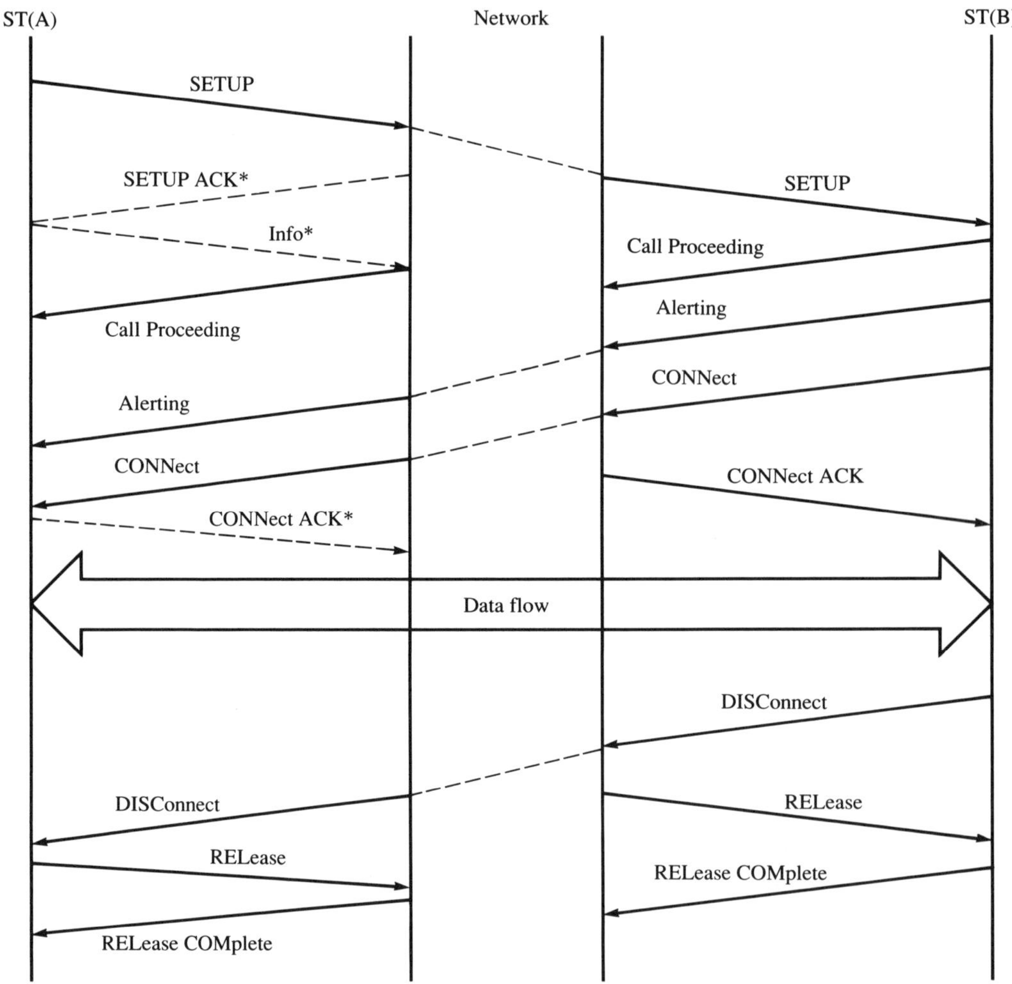

Messages with asterisks are optional depending on implementation.
ST: subscriber terminal

Figure III.1 Procedure for a simple circuit-switched call control

Table III.1 Message sizes and transmission delays for selected D-channel messages (cf. ECMA-106)

	Mandatory (bytes)		Optional[a] (bytes)		Simulation model	
Message	Min.	Max.	Min.	Max.	Nominal (bytes)	Tx[b] time (ms)
SETUP	10	19	22	169	20	2.5
CONN	3	6	18	137	5	0.625
CONN_ACK	3	6	3	34	5	0.625
DISC	7	42	7	68	10	1.25
REL	7	42	7	68	10	1.25
REL–COM	7	42	3	34	LI[c]	LI[c]

[a] Optional (min/max): if all the optional information elements are used.
[b] Tx: Transmission.
[c] LI: Applies to the local interface only in the simulation model.

Table III.2 Message sizes and transmission delays for selected D-channel messages (cf. Q.931)

	Mandatory (bytes)[a]		Optional (bytes)[b]	
Message	Min.	Max.	Min.	Max.
SETUP	8	17+	38	264+
CONN	4	4+	20	204+
CONN_ACK	4	4+	6	39+
DISC	8	36+	12	179+
REL	4	4+	12	207+
REL_COM	4	4+	12	207+

[a] Mandatory (max): values are network and service-dependent (for some information elements in messages).
[b] Optional (max): dependent upon the number and type of optional information elements used as well as being network and service-dependent.
[c] Fields are marked with '+'.

APPENDIX

IV

MULTI-LEVEL SERVICE RATE CONTROL

The steady-state system equations for a queuing system with a single queue and multiple identical parallel servers operating under a hysteresis threshold control have been derived by Moder and Phillips (1962). The queuing system is assumed to have Poisson arrivals with parameter λ and exponential service time distributions with parameter μ. It is further assumed that the switching times (channel-on or off) are zero and the number of servers is limited. A pair of thresholds of high (N, TH_H or H) and low (ν, TH_L or L) 'watermark' type are used over the queue length for controlling the service rate (or the number of active servers) of the queuing system.

The control doctrine operates as follows. Arrivals are queued and served under the strict first come–first served discipline. The system has a minimum of σ servers active even when the queue length is zero. The number of active servers is increased when the queue length reaches the threshold high from below. The number can thus be increased until it reaches an absolute maximum of S ($S > \sigma$) servers. The queue length is then unbounded. Servers are switched off when the queue length drops to ν ($0 \leq \upsilon \leq N-2$) and a service is completed.

This appendix re-examines the equations developed by Moder and Phillips and presents my own version with some further results. It is this version that has been used in this book.

IV.1 NOTATION

The following notation is used, as defined in Moder and Phillips (1962). The corresponding notation used in the rest of this book is also shown where applicable:

n =	number of customers in the queue (queue length)
s =	number of customers in service (i.e. number of busy channels)
L (N) =	mean number of customers in the system
L_q (Q) =	mean queue length

M = number of (fixed) channels in the conventional multiple channel process

N (TH_H or H) = threshold high; the 'shift-up' point of queue length at which additional channels are instantaneously switched in if $s < S$

v (TH_L or L) = threshold low; the 'shift-down' point, i.e. the minimum queue length, when $s > \sigma$

S = maximum number of channels, $\sigma < S < \infty$

σ = minimum number of channels, $\sigma \geq 1$

λ = mean arrival rate

μ = mean service rate per channel

ρ = λ/μ, the utilization factor (differs from the conventional multiple channel models where $\rho = (\lambda/M\mu)$)

γ = ρ/σ

$P_{n,s}$ = steady-state probability that n customers are in the queue and s units are being served

IV.2 SOLUTION OF MODEL EQUATIONS

It has been shown in Moder and Phillips (1962) that the solution of the model equations in terms of $P_{0,0}$ is as follows:

$$P_{0,s} = \left(\frac{\rho^s}{s!}\right) P_{0,0} \qquad (n=0;\ 0 \leq s \leq \sigma) \qquad \text{(i)}$$

$$P_{n,\sigma} = \gamma^n \left(\frac{\rho^\sigma}{\sigma!}\right) P_{0,0} \qquad (0 \leq n \leq v;\ s=\sigma) \qquad \text{(ii)}$$

$$P_{n,\sigma} = \left(\frac{\rho^\sigma}{\sigma!}\right)\left(\frac{\gamma^n - \gamma^N}{1-\gamma^{N-v}}\right) P_{0,0} \qquad (v \leq n \leq N-1;\ s=\sigma) \qquad \text{(iii)}$$

$$P_{n,s} = P_{v,s} \qquad (v \leq n \leq N-1;\ \sigma+1 \leq s \leq S-1) \qquad \text{(iv)}$$

$$P_{n,s} = \left(\frac{\rho^s}{s!}\right)\left(\frac{\gamma^{N-1} - \gamma^N}{1-\gamma^{N-v}}\right) P_{0,0} \qquad (v \leq n \leq N-1;\ \sigma+1 \leq s \leq S-1) \qquad \text{(v)}$$

$$P_{n,S} = \frac{\rho^S (S^{n-v+1} - \rho^{n-v+1})(\gamma^{N-1} - \gamma^N)}{S^{n-v} S! (S-\rho)(1-\gamma^{N-v})} P_{0,0} \qquad (v \leq n \leq N-1;\ s=S) \qquad \text{(vi)}$$

$$P_{n,S} = \frac{\rho^{S-N+n+1} (S^{N-v} - \rho^{N-v})(\gamma^{N-1} - \gamma^N)}{S^{n-v} S! (S-\rho)(1-\gamma^{N-v})} P_{0,0} \qquad (n > N-1;\ s=S) \qquad \text{(vii)}$$

The value of $P_{0,0}$ is given as:

$$P_{0,0} = \left\{ 1 + \sum_{s=1}^{\sigma} \frac{\rho^s}{s!} + \frac{\rho^\sigma \gamma}{\sigma!(1-\gamma)} - \frac{\rho^\sigma \gamma^N (N-v)}{\sigma!(1-\gamma^{N-v})} + \frac{(N-v)(\gamma^{N-1} - \gamma^N)}{1-\gamma^{N-v}} \left[\frac{\rho^S}{(S-1)!(S-\rho)} - \frac{\rho^\sigma}{\sigma!} + \sum_{s=\sigma}^{S-1} \frac{\rho^s}{s!} \right] \right\}^{-1}. \qquad \text{(viii)}$$

IV.3 MEAN NUMBER IN SYSTEM

The mean number of customers (packets) in the system, L, is the expected value of $(N+s)$, which can be written as follows (Moder and Phillips 1962):

$$L=\sum_{s=1}^{\sigma} sP_{0,s}+\sum_{n=1}^{v-1}(n+\sigma)P_{n,\sigma}+\sum_{n=v}^{N-1}(n+\sigma)P_{n,\sigma}+\sum_{n=v}^{N-1}\sum_{s=\sigma+1}^{S-1}(n+s)P_{n,s}+\sum_{n=v}^{N-2}(n+S)P_{n,S}+\sum_{n=N-1}^{\infty}(n+S)P_{n,S}\,. \tag{IV.1}$$

Equation (IV.1) gives the following closed form expression for L:

$$L=\frac{P_{0,0}}{(1-\gamma^{N-v})}\left((1-\gamma^{N-v})\left\{\rho\sum_{s=0}^{\sigma-1}\frac{\rho^s}{s!}+\frac{\rho^\sigma}{\sigma!}\left[\frac{\gamma-v\gamma^v+(v-1)\gamma^{v+1}}{(1-\gamma)^2}+\frac{\sigma(\gamma-\gamma^v)}{1-\gamma}\right]\right\}\right.$$

$$+\frac{\rho^\sigma}{\sigma!}\left\{\frac{(N-1)\gamma^{N+1}-N\gamma^N-(v-1)\gamma^{v+1}+v\gamma^v}{(1-\gamma)^2}+\frac{\sigma(\gamma^v-\gamma N)}{1-\gamma}-\frac{1}{2}\gamma^N\left[N(N-1)-v(v-1)\right]-\sigma\gamma^N(N-v)\right\}$$

$$+\frac{1}{2}\gamma^{N-1}(1-\gamma)\left\{\left[N(N-1)-v(v-1)\right]\sum_{s=\sigma+1}^{S-1}\frac{\rho^s}{s!}+2(N-v)\rho\sum_{s=\sigma}^{S-2}\frac{\rho^s}{s!}\right\}+\frac{\rho^S\gamma^{N-1}(1-\gamma)}{S!(S-\rho)}$$

$$\times\left\{\frac{1}{2}S(N-1)(N-2)+S^2(N-v-1)-\frac{(N-2)\rho^{N-v+1}-(N-1)S\rho^{N-v}+\rho^2S^{N-v-1}}{S^{N-v-2}(S-\rho)^2}-\frac{\rho S^{N-v-1}-\rho^{N-v}}{S^{N-3-v}(S-\rho)}\right.$$

$$\left.-\frac{1}{2}Sv(v-1)-\frac{v\rho S}{S-\rho}\right\}$$

$$\left.+\frac{\gamma^{N-1}(1-\gamma)(S^{N-v}-\rho^{N-v})\rho^S}{S!(S-\rho)^3S^{N-v-2}}\left[S(N-1)-\rho(N-2)+S(S-\rho)\right]\right) \tag{IV.2}$$

This is different in a few terms from the equation for L quoted in the paper by Moder and Phillips (1962).

The plot of (IV.2) for values of $(0 \le \rho < 10)$ with a setting of minimum number of servers allowed to be 1 and maximum number allowed to be 10 is depicted in Fig. IV.1. The graph shows multiple plots for different values of N (threshold high, H) while keeping the threshold low (L) value constant at zero. These plots show a characteristic behaviour. A 'knee' occurs at the critical value of $\rho=1$. From here on until relatively high values of 9, the curves are increasing with a positive slope. At that second critical point, the queuing system is operating at very high traffic intensities and the queue starts to build up. The reason for the 'knee' at 1 is that the system tries to optimize the usage of servers and in doing so switches only as many servers as necessary to limit the queue build-up; as the switching of the second (extra) server is delayed, the knee forms.

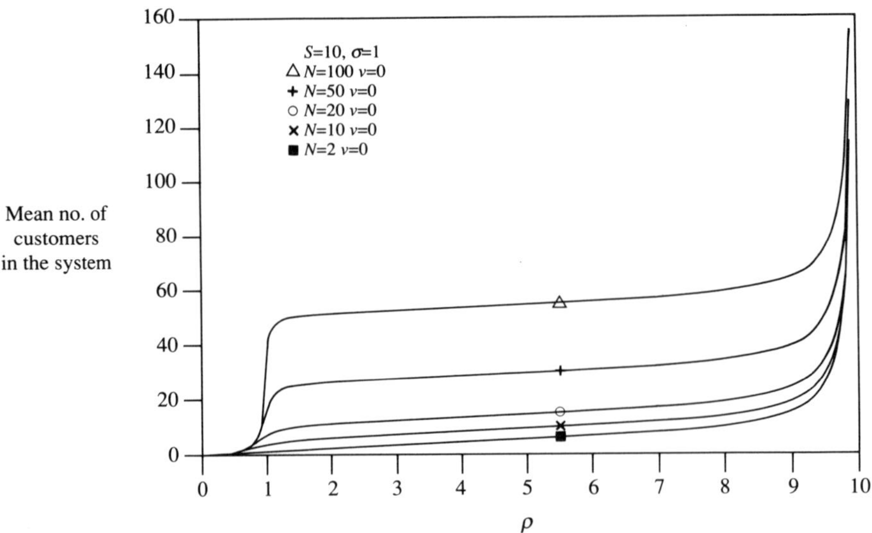

Figure IV.1 Mean number of customers *v*. utilization factor (ρ) (equation (IV.2))

A further check for the system is to substitute the values obtained for L and L_q into an equation relating the two measures. This is shown below.

IV.4 MEAN QUEUE LENGTH

The mean queue length L_q is given by the following equation (Moder and Phillips 1962):

$$L_q = \sum_{s=0}^{\sigma}(0)P_{0,s} + \sum_{n=1}^{\nu} nP_{n,\sigma} + \sum_{n=\nu+1}^{N-1} nP_{n,\sigma} + \sum_{n=\nu}^{N-1}\sum_{s=\sigma+1}^{S-1} nP_{n,s} + \sum_{n=\nu}^{N-2} nP_{n,S} + \sum_{n=N-1}^{\infty} nP_{n,S} \qquad \text{(IV.3)}$$

This can be shown to give the following closed-form expression for L_q:

$$L_q = \frac{P_{0,0}}{(1-\gamma^{N-\nu})}\left(\frac{\rho^{\sigma}[\,\gamma-(\nu+1)\gamma^{\nu+1}+\nu\gamma^{\nu+2}\,](1-\gamma^{N-\nu})}{\sigma!(1-\gamma)^2}\right.$$

$$+\frac{\rho^{\sigma}}{\sigma!}\left\{\frac{\gamma^{N}(N\gamma-\gamma-N)+\gamma^{\nu+1}(\nu+1-\nu\gamma)}{(1-\gamma)^2}-\frac{1}{2}\gamma^{N}\left[N(N-1)-\nu(\nu+1)\right]\right\}$$

$$+\frac{1}{2}\gamma^{N-1}(1-\gamma)\left[N(N-1)-\nu(\nu-1)\right]\left\{\sum_{s=\sigma}^{S-1}\frac{\rho^{s}}{s!}-\frac{\rho^{\sigma}}{\sigma!}\right\}$$

$$+\frac{\rho^{S}\gamma^{N-1}(1-\gamma)}{2S!(S-\rho)^3S^{N-2}}\left\{S^{N-1}(S-\rho)^2\left[(N-1)(N-2)-\nu(\nu-1)\right]\right.$$

$$-2S^{v}\rho^{N-v}\left[\rho(N-2)-S(N-1)\right]-2S^{N-v-1}\rho^{v+1}\left[Sv-(v-1)\rho\right]+2(S^{N-v}-\rho^{N-v})S^{v}\left[S(N-1)-\rho(N-2)\right]\Bigg\}\Bigg) \quad \text{(IV.4)}$$

IV.5 RESULTS

Tables IV.1 and IV.2 give the values obtained from (IV.2) and (IV.5) with the following settings: v (TH_L)=0, S=2 and S=3 for ρ=0.5 and ρ=1.5, respectively.

Table IV.1 Results when $\rho = 0.5$ ((IV.2) and (IV.5))

	S=2			S=3		
Measure	N=2	N=4	N=8	N=2	N=4	N=8
Lq	0.1458	0.3217	0.4693	0.1245	0.3119	0.4677
L	0.6458	0.8217	0.9693	0.6245	0.8119	0.9677

Table IV.2 Results when $\rho = 1.5$ ((IV.2) and (IV.5))

	S=2			S=3		
Measure	N=2	N=4	N=8	N=2	N=4	N=8
Lq	2.2500	3.0036	4.7399	0.5902	1.3823	3.1502
L	3.7500	4.5036	6.2399	2.0902	2.8823	4.6502

It is also known that, for an M/M/c queuing system with identical servers, the following equation holds true (Gross and Harris 1985):

$$L = L_q + \left(\frac{\lambda}{\mu}\right) \quad \text{(IV.5)}$$

where $\rho = \lambda / \mu$ is the traffic intensity for one server. Therefore it can be seen that the tables adhere to equation (IV.5).

IV.6 SUMMARY

Equations for L and L_q have been presented. A plot of the equation for the number in the system versus the traffic intensity was also given, providing further insight into the system behaviour at relatively large values of the input parameters. A modification to the equation presented in Moder and Phillips (1962) for L removes the negative values obtained. It is this modification that has been used throughout this book.

APPENDIX

V

DELAYED CLOSE POLICY

This appendix presents a routine for delayed channel close policy (DCP). This policy can typically have two variants. Type 1 (*DCP1*) prevents the reduction of the channel bandwidth when the time duration elapsed since the last bandwidth increase (the assumed epoch where the superchannel charging period is restarted) is less then n per cent of the charged unit time. Policy type 2 (*DCP2*) operates with the normal bandwidth management policy with all channel bandwidth values except the last BBU. That is, the DCP comes into play whenever the server capacity is reduced to 1 BBU so that it will be reluctant to shut down a channel (server) completely until such time as no new packet arrivals are recorded within n per cent of the charged unit time. This ensures that at least one server is available at all times from the time of last packet arrival to m per cent of the charged unit time.

Assuming that the condition for bandwidth reduction has occurred,[1] the delayed channel close routine is called. Figure V.1 shows simple pseudo-code descriptions of the two policies DCP1 and DCP2. In both cases, the initiation of the delayed close causes the calculation of the time τ, when the connect_time will be within n per cent of the charged unit time. This is then queued as a self-timed event, *dclose_event*, to occur at τ and has got all the information about the channel to be closed (or bandwidth decremented). When dclose_event matures, a check is made to see if a packet arrival has occurred since the initiation of the delayed close operation. If none has arrived, the bandwidth decrement/channel close is processed; otherwise delayed close is aborted.

[1] Each *packet sent* message causes the channel management routine to check the *qsize* variable holding that particular channel's queue size (or number of packets in the system). This could then trigger a bandwidth reduction condition if the value in the *qsize* variable is equal to or lower than the lower threshold of a hysteresis service capacity control policy.

DCP 1:

if { (bw > 1) AND (connect_time within n% of *charged unit time*) }
   then DECREMENT channel bandwidth by 1 BBU
else if { (bw == 1) AND (connect_time within n% of *charged unit time*) }
        then CLOSE channel
     else INITIATE *Delayed Close*

DCP 2:

if { (bw > 1)
   then DECREMENT channel bandwidth by 1 BBU
else if { (bw == 1) AND (connect_time within n% of *charged unit time*) }
        then CLOSE channel
     else INITIATE *Delayed Close*

Figure V.1 Delayed close policies (DCPs) 1 and 2

APPENDIX

VI

ACRONYMS AND ABBREVIATIONS

ACE	automatic cross connection equipment
ACK	acknowledgement
ACT./Act.	activity
Add./addr.	address
ANSI	American National Standards Institute
API	application program interface
APMBS	additional packet mode bearer service
ARP	address resolution protocol (process)
ATM	asynchronous transfer mode
AU	access unit
B-channel	64 kbps channel
BA	bandwidth allocation
BAP	bandwidth allocation policy
BAS	bit-rate allocation signal
BBU	basic bandwidth unit
BCP	B-channel process
BER	bit error rate
BISDN/B–ISDN	broadband ISDN
blk.	block
BM	bandwidth management
BMP	bandwidth management policy
BMU	bandwidth management unit
bps	bits per second
Bps	bytes per second
BRA	basic-rate access
BRISDN	basic-rate ISDN
BW	bandwidth

CAU	channel allocation unit
CCB	channel control message block
CCIR	International Radio Consultative Committee
CCITT	The International Telegraph and Telephone Consultative Committee
CCS	common channel signalling
CCST	circuit control status table
cct	circuit
CEC	Commission of European Communities
CEN/CENELEC	European Committee for Standardisation/European Committee for Electrotechnical Standardisation
CEPT	European Conference of Postal & Telecommunications Administrations (Conference Europeénne des Administrations des Postes et des Télécommunications)
CH./ch.	channel
CHBW	change bandwidth
ChP	charge period
CI	confidence interval
CIIE	channel identification information element
CL	connectionless
CLNS	connectionless network service
CLTP	connectionless transport protocol
CM	channel management
CMA	channel management architecture
CMFU	channel management functional unit
CMP	channel management policies
CMR	channel management routine
CO	connection oriented (connection mode)
CONN	connected
CONS	connection oriented network service
Cons.	connections
COTP	connection oriented transport protocol
CP	conversion protocol
CPE	customer premises equipment
CQ	channel queue
CRIE	call reference information element
CRN	call reference number
CS	circuit switching
CS-BS	circuit-switched bearer service
CSMA/CD	carrier sense multiple access with collision detection
CSPDN/CSDN	circuit-switched public data network
CST	channel status table
DASS-2	digital access signalling system version 2
DCCP	delayed channel close policy
DCM	dynamic channel management
DCMA	dynamic channel management architecture
DCO	data communications oriented

DCP	D-channel process
DEC.	decrement
DECR_BW	decrement bandwidth
DISC.	disconnect
DLL	data link layer
DMA	direct memory access
DPM	dual ported memory
DPNSS	digital private network signalling system
DQDB	distributed queue dual bus
Dst./dst.	destination
DTE	data terminal equipment
DTI	Department of Trade and Industry (UK)
DTP	distributed transaction processing
E-mail	electronic mail
ECMA	European Computer Manufacturers Association
ECSA	Exchange Carriers Standards Committee (USA)
EIR	ethernet ISDN relay
ES	end system
ESPRIT-I	European Strategic Programme of Research in Information Technology
ET	exchange termination
ETS/NET	European Telecommunication Standards
ETSI	European Telecommunications Standards Institute
FAS	frame alignment sequence
FDDI	fiber distributed data interchange
HDLC	high-level data link control
HLD	high-level decision
HLDP	high-level decision process
HP	host processor
HPB	host processor board
HRX	hypothetical reference connections
IAT	(packet) inter arrival time
IcLAN	ISDN compatible LAN
IDA	integrated digital access
IDN	integrated digital network
IDP	ISDN interface driver process
IEEE	Institute of Electrical and Electronic Engineering (USA)
IFU	interworking functional unit
IICMP	ISDN interface control main process
IIM	ISDN interface module
INC.	increment
INCA	integrated network communications architecture
io.	input/output
io_blk_p	input/output block pointer
IP	(DARPA) internet protocol
ipdst.	IP destination
IPRP	IP routing process

ipsrc.	IP (Internet Protocol) source
IRP	ISDN router (routing) process
IS	intermediate system
ISDN	integrated digital services network
Isdst/isdst.	ISDN destination
ISIM	ethernet ISDN relay and ISDN simulator
ISLAN	integrated services local area network
ISLN	integrated services local network
ISN	integrated services network
ISO	International Standards Organisation
ISO/IEC	ISO/International Electrotechnical Commission
ISO/JTC	ISO/Joint Technical Committee
ISPBX/ISBX	integrated services private branch exchange
Issrc	ISDN source
ITSTC	CEN/CENELEC Steering Committee for Information Technology
ITSU	Information Technology Standards Unit (DTI)
IU/IWU	interworking unit
IVD	integrated voice data
JANET	Joint Academic Network (UK)
LAN	local area network
LAP-B	link access protocol balanced
LAP-D	link access protocol for the D-channel
LC	logical channel
LCRN	logical channel reference number
LC_BW	logical channel bandwidth
LLC	logical link control
LLC-IE	lower layer compatibility information element
LLCIP	low-level control interface process
LLD	low level decision
LLDP	low level decision process
M	mega (10^6)
MAC	media access control
MAN	metropolitan area network
MI	management information
MLP	multi-link procedure (protocol)
MS-LAN	multi-service LAN
Msg./msg.	message
MSL	maximum segment lifetime
NB	narrow band (width)
NET/ETS	Normes Européennes de Télécommunication
NIST	National Institute of Standards and Technology (USA)
NIU Forum	North American ISDN Users' Forum
NL	network layer
NPDU	network protocol data unit
NS	network service
NSAP	network service access point

NT	network termination
OSI	open systems interconnection
OSI-NS	OSI network service
OSI-RM	OSI reference model
PABX	private automatic branch exchange
PBX	private branch exchange
PC	port command
PCB	port control block shared data structure
PCM	pulse code modulation
PCR	port control register
PCRN	physical channel number
PD	port driver
PDU	protocol data unit
PGEN	packet generator
PH	packet handler
PIB	port interface board
PICC	port interface communications and control unit
PIP/PP	port (interface) processor
PKT/pkt.	packet
PL	physical layer
PLP	packet level protocol (X.25 Layer 3)
PRA	primary-rate access
PRBS	pseudo-random binary sequence
PRISDN	primary-rate ISDN
PROOF	primary-rate OSI office
PS-BS	packet-switched bearer service
PSINK	packet sink
PSIZE	packet size
PSPDN	public switched packet data network
PSTN	public switched telephone network
PTT	post telephone telegraph
PVC	permanent virtual circuit
P_CH	pending channel
QLEN/Qlen/qlen	queue length
QoS	quality of service
qsize	queue size
R	reference point in ISDN local access arrangement
RA	restricted access
Recs.	recommendations
Ref./ref.	reference
REL	release
RESV./resv.	reserved
RLM	relay module
RP	routing policy
RT	response time (sojourn time)
RT	routing table

RTO	retransmission time-out
Rx	receive
RxQ	receive queue
S	reference point in ISDN local access arrangement
SAP	service access point
SAPI	service access point identifier
SB	status block
sb_p	status block pointer
SC/S_ch	superchannel
SDF	synchronous data formatter
SENET	slotted envelope network
SI_PDU	superchannel identification protocol data unit
SLOANE	satellite and local area network environment
SLP	single link procedure (protocol)
SNAcP	sub-network access protocol
SNDCP	sub-network dependent convergence protocol
SNICP	sub-network independent convergence protocol
SPAG	standards promotion and application group
SRC	service rate control
SSN 7	signalling system number 7
Sta.	state
STM	synchronous transfer mode
T	reference point in ISDN local access arrangement
TA	terminal adaptor
TCO	telecommunications oriented
TCP	(DARPA) transport control protocol
TDM	time division multiplex(ing)
TDMA	time division multiple access
TE	terminal equipment
TH_H/H	threshold high
TH_L/L	threshold low
TLR	transport layer relay
TO/ T/O	time out
TP	transport protocol
TPDU	transport protocol data unit
TS	time slot
TSMAP	time slot mapping
TSSI	time slot sequence integrity
TWTO	time wait time-out
Tx	transmit
TxQ	transmit queue
U	user equipment
UCL	University College London
UE	User Equipment
UISF	UK ISDN Standards Forum
UM	user module

UNI	user–network interface
UTO	user time-out
UUIE	U–U information element
UUS	user–user signalling
VAS	valued added service
VC	virtual circuit
VLSI	very large scale integration (or integrated circuits)
WAN	wide area network
WB	wide band(width)
WS	workstation

APPENDIX

VII

CCITT RECOMMENDATIONS ON ISDN

Several series of CCITT Recommendations relate, or have sections that relate, to ISDN. These are the I, Q, X, V, E, T and G-Series of Recommendations. The Recommendations of the I-Series are the most extensive ones covering many aspects of ISDNs. These are presented in detail below. Others are given in broader terms.

THE CCITT I-SERIES RECOMMENDATIONS[1]

Part I: General Structure

Section 1: Framework of I-Series Recommendations

I.110 Preamble and General Structure of the I-Series Recommendations
I.111 Relationship with Other Recommendations Relevant to ISDNs
I.112 Vocabulary of Terms for ISDNs
I.113 Vocabulary of Terms for Broadband Aspects of ISDN

Section 2: Description of ISDNs

I.120 Integrated Services Digital Networks
I.121 Broadband Aspects of ISDN
I.122 Framework for Providing Additional Packet-Mode Bearer Services

[1] The list given here is based on the Blue Book (CCITT 1988a). The CCITT in its IXth Plenary Assembly in Melbourne decided that Recommendations can be adopted by correspondence as and when prepared by the Study Groups. Some interim Recommendations have also been adopted since the IXth Plenary Assembly; these are indicated by an asterisk next to them. Some other recommendations have been revised; these are indicated by an ‡ placed next to them. The most recent Plenary Assembly took place in Helsinki in March 1993, when further Recommendations were adopted.

Section 3: General Modelling Methods

I.130 Method for the Characterisation of Telecommunication Services Supported by an ISDN and Network Capabilities of an ISDN

Section 4: Telecommunication Network and Service Attributes

I.140 Attribute Technique for the Characterization of Telecommunication Services supported by an ISDN and Network Capabilities of an ISDN
I.141 ISDN Network Charging Capabilities Attributes

Part II: Service Recommendations

I.200 Guidance to the I.200 Series of Recommendations

Section 1: General Aspects of Service in ISDN

I.210 Principles of Telecommunication Services supported by an ISDN and the Means to Describe Them

Section 2: Common Aspects of Services in the ISDN

I.220 Common Dynamic Description of Basic Telecommunications Services
I.221 Common Specific Characteristics of Services

Section 3: Bearer Services supported by an ISDN

I.230 Definition of Bearer Service Categories
I.231 Circuit-Mode Bearer Service Categories
I.232 Packet-Mode Bearer Service Categories
I.2xy* ISDN Frame Mode Bearer Services

Section 4: Teleservices supported by an ISDN

I.240 Definition of Teleservices
I.241 Teleservices supported by an ISDN

Section 5: Supplementary Services in ISDN

I.250 Definition of Supplementary Services
I.251 Number Identification Supplementary Services
I.251.8* Sub-Addressing Supplementary Service
I.252 Call Offering Supplementary Services
I.253 Call Completion Supplementary Services
I.254 Multiparty Supplementary Services
I.255 Community of Interest Supplementary Services
I.255.1* Closed User Group (CUG)
I.256 Charging Supplementary Services
I.257 Additional Information Transfer
I.257.1* User-to-User Signalling

Part III: Overall Network Aspects and Functions

Section 1: Network Functional Principles

I.310 ISDN Network Functional Principles

Section 2: Reference Models

I.320‡ ISDN Protocol Reference Model
I.324‡ ISDN Network Architecture
I.325‡ Reference Configurations for ISDN Connection Types
I.326 Reference Configurations for Relative Network Resource Requirements

Section 3: Numbering, Addressing, and Routing

I.330 ISDN Numbering and Addressing Principles
I.331 Number Plan for the ISDN Era
I.332 Numbering Principles for Interworking between ISDNs and Dedicated Networks with Different Numbering Plans
I.333‡ Terminal Selection in ISDN
I.334 Principles Relating ISDN Numbers/Subaddresses to the OSI Reference Model Network Layer Addresses
I.335 ISDN Routing Principles

Section 4: Connection Types

I.340 ISDN Connection Types

Section 5: Performance Objectives

I.350 General Aspects of Quality of Service and Network Performance in Digital Networks, including ISDN
I.351‡ Recommendations in Other Series Concerning Network Performance Objectives that Apply at Reference Point T of an ISDN
I.352 Network Performance Objectives for Connection Processing Delays in an ISDN
I.35a* Availability Performance for 64 kbps ISDN Connection Types
I.35e* Reference Events for Defining ISDN Performance Parameters
I.35p* Network Performance Objectives for Packet-Mode Communication in an ISDN
I.3xx* Congestion Management for the Frame Relaying Bearer Service

Part IV: ISDN User–Network Interfaces

Section 1: ISDN User–Network Interfaces

I.410 General Aspects and Principles Relating to ISDN User–Network Interfaces
I.411 ISDN User–Network Interfaces: Reference Configurations
I.412 ISDN User–Network Interfaces: Interface Structures and Access Capabilities

Section 2: Application of I-Series Recommendations to ISDN User–Network Interfaces

I.420 Basic User–Network Interface
I.421 Primary Rate User–Network Interface

Section 3: ISDN User–Network Interfaces: Layer 1 Recommendations

I.430 Basic User–Network Interface: Layer 1 Specification
I.431 Primary-Rate User–Network Interface: Layer 1 Specification

Section 4: ISDN User–Network Interfaces: Layer 2 Recommendations

I.440 ISDN User–Network Interface Data Link Layer: General Aspects
I.441 ISDN User–Network Interface Data Link Layer: Specification

Section 5: ISDN User–Network Interfaces: Layer 3 Recommendations

I.450 ISDN User–Network Interface Layer 3: General Aspects
I.451 ISDN User–Network Interface Layer 3: Specification for Basic Call Control
I.452 Generic Procedures for the Control of ISDN Supplementary Services

Section 6: Multiplexing, Rate Adaptation, and Support of Existing Interfaces

I.460 Multiplexing, Rate Adaptation, and Support of Existing Interfaces
I.461 Support of X.21-, X.21 bis- and X.20 bis-based DTEs by an ISDN
I.462 Support of Packet Mode Terminal Equipment by an ISDN
I.463 Support of DTEs with V-Series Type Interfaces by an ISDN
I.464‡ Multiplexing, Rate Adaptation and Support of Existing Interfaces for Restricted 64 kbps Transfer Capability
I.465 Support by an ISDN of DTEs with V-Series Type Interfaces with Provisions for Statistical Multiplexing

Section 7: Aspects of ISDN Affecting Terminal Requirements

I.470 Relationship of Terminal Functions to ISDN

Part V: Internetwork Interfaces

I.500 General Structure of ISDN Interworking Recommendations
I.510 Definitions and General Principles of ISDN Interworking
I.511 ISDN-to-ISDN Layer 1 Internetwork Interface
I.515‡ Parameter Exchange for ISDN Interworking
I.520‡ General Arrangements for Network Interworking between ISDNs
I.530‡ Network Interworking between and ISDN and a PSTN
I.540 General Arrangements for Interworking between CSPDNs and ISDNs for the Provision of Data Transmission
I.550 General Arrangements for Interworking between PSPDNs and ISDNs for the Provision of Data Transmission
I.560 Requirements to be Met in Providing the Telex Service within the ISDN
I.5xz* Frame Mode Bearer Service Interworking

Part VI: Maintenance Principles

I.601 General Maintenance Principles of ISDN Subscriber Access and Subscriber Installation

I.602 Application of Maintenance Principles to ISDN Subscriber Installation
I.603 Application of Maintenance Principles to ISDN Basic Accesses
I.604 Application of Maintenance Principles to ISDN Primary-Rate Accesses
I.605 Application of Maintenance Principles to Static Multiplexed ISDN Basic Accesses

Q-SERIES RECOMMENDATIONS

Q.500 Digital Exchanges
Q.700 Interexchange Signalling (Signalling System no. 7)
Q.920–930 User–Network Interface (correspond to I.440–450 Series)
Q.71–99 and Q.937 Supplementary Services

X-SERIES RECOMMENDATIONS

X.30–31 Support of X-Series Data Terminals by an ISDN

V-SERIES RECOMMENDATIONS

V.110 Support of V-Series Data Terminals by an ISDN
V.120 Support of V-Series Data Terminals with Provision for Statistical Multiplexing by an ISDN

G-SERIES RECOMMENDATIONS

G.700, G.800, G.900 Series Digital Transmission

E-SERIES RECOMMENDATIONS

E.164 Numbering Plan for the ISDN Era (corresponds to I.331)
E.165 Timetable for Implementation of E.164
E.166 Numbering Plan Interworking in the ISDN Era
E.167 ISDN Network Identification Codes
E.172 Call Routing in the ISDN Era
E.184 Indications to Users of ISDN Terminals
E.213, E-215, E.216 Maritime Mobile Service and ISDN

T-SERIES RECOMMENDATIONS

T.90 ISDN Telematic Terminals

REFERENCES

Adams, J.L. and Falconer, R.M. (1984). Orwell: a protocol for carrying integrated services on a digital communications ring. *Electronics Letters*, **20**, 970–1.

Altarah, L. and Motard, S. (1989). Dynamic use of ISDN B-channels. Paper presented at the ISDN in Europe '89, The Hague, 25–27 April 1989.

Altarah, L. and Seret, D. (1991). Dynamic management of B-channels for Nx64 kbit/s LANs interconnection through the ISDN. Paper presented at the 2nd International Conference on Local Communication Systems: LAN and PBX, Palma de Mallorca, 26–28 June 1991.

Appenzeller, H.R. (1986). Signalling system no. 7 ISDN user part. *IEEE JSAC*, **4**(3), 366–71.

AT&T (1989). *T7115 SPYDER: Synchronous Data Formatter*. AT&T.

Atkins, J.W. and Cooper, N.J.P. (1989). The development of packet data communications in the ISDN. Paper presented at the Second IEE National Conference on Telecommunications, York (UK), 2–5 April 1989.

Standards Australia (1991). Information processing: digital channel aggregation (Nx64). Draft Australian Standard (DR 91140 : R) (1 June).

Banks, J. and Carson, J.S. (1984). *Discrete-Event System Simulation*, Prentice-Hall, Englewood Cliffs, NJ.

Baskett, F., Chandy, K.M., Muntz, M.M. and Palacios, F.G. (1975). Open, closed and mixed networks of queues with different classes of customers. *J. of the ACM*, **22**(2), 248–60.

Bell, C.E. (1980). Optimal operation of an M/M/2 queue with removable servers. *OR*, **28**(5), 1189–1204.

Berera, E. and Jardin, P. (1987). Interconnection of LANs and PABXs in an ISDN environment. Paper presented at the ISDN'87, London, 16–18 June 1987.

Bhattacharya, P.P., Viniotis, I., and Ephremides, A. (1987). A (not very) simple routing problem. Paper presented at the 25th Allerton Conference on Communications, Control and Computing, University of Illinois at Urbana-Champaign, 30 September–2 October 1987.

Boeing (1988). *TOP: Technical and Office Protocols Specification (v.3.0)*. Boeing Corporation.

Boltz, B., Liang, Y., and Roberts, J. (1989). Provision of Nx64 Kbit/s services in the ISDN. Paper presented at the ISDN in Europe '89, The Hague, 25–27 April 1989.

Bonding Consortium (1992). Interoperability Requirements for Nx56/64 kbps Calls, (Voting Draft V1.0), 15 May 1992.

Bratley, P., Fox, B.L. and Schrage, L.E. (1983). *A Guide to Simulation*. Springer-Verlag, New York City.

Brill, P.H. and Green, L. (1984). Queues in which customers receive simultaneous service from a random number of servers: a system point approach. *Management Sci.*, **30**(1), 51–68.

British Telecom (1985). Digital Access Signalling System no. 2 (DASS-2). *BTNR*, 190(1).

British Telecom (1989). The Digital Private Network Signalling System (DPNSS). *BTNR*, 188(5).

Buhler, H. (1982). *Reglages echantillonnes*. Presses polytechniques romandes, Paris.

Burg, F.M. and Iorio, N.D. (1989). Networking of networks: interworking according to OSI. *JSAC*, **7**(7), 1131–42.

Burren, J.W. (1985). Project universe: overview of the project. Project Report no. 1. Rutherford Appleton Laboratory, Oxford.

Burren, J.W. (1989). Flexible aggregation of channel bandwidth in primary rate ISDN. Paper presented at SIGCOMM'89, Austin, Texas, 19–22 September 1989.

CCITT (1984). *Recommendations of the I-Series* (Red Book). Malaga–Torremolinos.

CCITT (1988a). *Recommendations of the I-Series ISDN* (Blue Book). Melbourne.

CCITT (1988b). *Basic User–Network Interface: Layer 1 Specification,* (Blue Book Vol. III, Fascicle III.8, Rec. I.430). Melbourne.

CCITT (1988c). *Primary Rate User–Network Interface: Layer 1 Specification,* (Blue Book Vol. III, Fascicle III.8, Rec. I.431). Melbourne.

CCITT (1988d). *Network Service Definition for Open Systems Interconnection for CCITT Applications* (Rec. X.213).

CCITT (1988e). *Use of X.25 to Provide the OSI Connection Mode Network Service for CCITT Applications* (Rec. X.223).

CCITT (1988f). *General Principles and Arrangements for Interworking between Public Data Networks, and between Public Data Networks and Other Public Networks* (Rec. X.300).

CCITT (1990). *Recommendation H.221: Line Transmission of Non-Telephone Signals — Frame Structure for a 64 to 1920 kbp/s Channel in Audiovisual Teleservices.*

Cerf, V.G. and Kirstein, P.T. (1978). Issues in packet network interconnection. *Proceedings of the IEEE*, **66**(11), 1386–1408.

Clark, P. (1986). Primary rate LAN interface for office applications. Paper presented at the ISDN Conference, **1**, 259–68 (Online Computer Communications Series 4), London.

Clark, P.F., Burren, J.W., Griffiths, J.W.R., Hinchley, A.J. and Needham, R.M. (1986). Unison: communications research for office applications. *Electronics and Power (IEE)*, **32**(9), 665–8.

CN&IS (1991). Memorandum of Understanding for the Implementation of a European ISDN Service by 1992. *CN&IS*, **21**(1), 69–74.

Cooper, M.D. (1987). The impact of ISDN on local area networks. Paper presented at the ISDN Conference 16–18 June 1987, London.

Coviello, G.J. and Vena, P.A. (1975). Integration of circuit/packet switching by a SENET concept. Paper presented to the National Telecommunciations Conference, New Orleans, 1-3 December 1975.

Crabill, T.B., Gross, D. and Magazine, M.J. (1977). A classified bibliography of research on optimal design and control of queues. *OR*, **25**(2), 219–32.

Daigle, J.N. and Whitehead, D. (1984). Transmission facility sharing for integrated services. Paper presented to the ICC'84 Conference, Amsterdam, 14-17 May 1984.

Davies, D.R. (1985). LANs and ISDN. Paper presented to the IEE International Conference on ISDN, 14–16 January 1985, London.

Day, J.D. and Zimmermann, H. (1983). OSI reference model. *Proc. of IEEE*, **71**(12), 1334–40.

Deniz, D.Z. (1985a). ISDN. In UCL Internal Note, 1839.

Deniz, D.Z. (1985b). ISLN. In UCL Internal Note, 1840.

Deniz, D.Z. (1986). Annual Research Report—Year 2 (UCL End-of-Year Report).

Deniz, D.Z. (1987a). ISDN—Userboard kit (Siemens) in the context of UCL ISDN test-bed studies. In UCL Internal Note 2115.

Deniz, D.Z. (1987b). UCL ISDN test-bed. In UCL Internal Note 2116.

Deniz, D.Z. (1991). Dynamic channel management in packet mode usage of circuit switched ISDN. Ph.D. thesis, University of London.

Deniz, D.Z. and Knight, G.J. (1989a). Channel management issues in LAN–ISDN interconnection. Proceedings of the ISDN in Europe Conference, Elsevier, North-Holland.

Deniz, D.Z. and Knight, G.J. (1989b). Managing multiple channels in LAN–ISDN interconnection. Paper presented to the ISCIS' IV, Cesme, Izmir, Turkey, 30 October–1 November 1989.

Deniz, D.Z. and Knight, G.J. (1990). On the formation and dynamic management of superchannels in primary rate ISDN. Proceedings of IFIP International Symposium on the Management of Local Communications Systems, Elsevier, North-Holland.

Deniz, D.Z. and Matthewson, P.W. (1986). Building wide area networks with the ISDN. In UCL Internal Note 1973.

DePrycker, M. (1985). LANs in an ISDN: consistency with the OSI reference model. *Computer Communications*, **8**(2), 74–8.

Du, X. and Knight, G. (1991). Further study of channel management issues in a LAN–ISDN gateway. UCL Research Note : RN/91/78 (June).

Duc, N.Q. and Chew, E.K. (1985). ISDN protocol architecture. *IEEE Communications*, **23**(3), 15–22.

ECMA (1985a). Physical layer at the primary rate access interface between data processing equipment and private circuit switching networks, ECMA-104 (September).

ECMA (1985b). *Layer 3 protocol for signalling over the D-channel of the S-interfaces between data processing equipment and private circuit switching networks*, ECMA-106 (September).

ECMA (1990a). Data link layer protocol for the D-channel of the interfaces at the reference point between terrminal equipment and private telecommunication networks, 3rd edn. ECMA-105 (June).

ECMA (1990b). Draft TR-LII: LAN interworking with ISDN and PTNX, ECMA/TC32-TG10/90 (October).

Ephremides, A., Varaiya, P., and Walrand, J. (1980). A simple dynamic routing problem. *IEEE Trans. on Automatic Control*, **AC-25**(4), 690–3.

Falconer, R.M. and Adams, J.L. (1985). Orwell: a protocol for integrated services local network. *BT Technol. J.*, **3**(4), 27–35.

Fan, J. (1989). Investigation into interconnecting the ISDN and existing packet switched networks and the experiments in the context of the UK interim ISDN–IDA. M.Phil. dissertation, University of London.

Fletcher, G.Y., Perros, H.G. and Stewart, W.J. (1986). A queueing network model of a circuit switching access scheme in an integrated services environment. *IEEE Trans. on Comms.*, **COM-34**(1), 25–30.

Frantzen, V. (1987). Trends in the development of public telecommunication networks. *Computer Networks and ISDN Systems*, **14**(2-5), 339–58.

Gagliardi, D. (1989). The role of ETSI in the European standardisation. Paper presented to the ISDN in Europe '89, The Hague, 25–27 April 1989.

Gebhard, R.F. (1967). A queuing process with bilevel hysteretic service-rate control. *NRLQ*, **14**(1), 55–68.

Gelenbe, E. and Mitrani, I. (1980). *Analysis and Synthesis of Computer Networks*, Academic Press, London.

Gelenbe, E. and Pujolle, G. (1976). The behaviour of a single queue in a general queuing network. *Acta Informatica*, **7**, 123–36.

Gelenbe, E. and Pujolle, G. (1987). *Introduction to queuing networks*. John Wiley, Chichester.

General Motors (1988). *MAP: Manufacturing Automation Protocols Specification (v.3.0)*. General Motors, Detroit.

Gitman, I., Hsieh, W-N, and Occhiogrosso, B.J. (1981). Analysis and design of hybrid switching networks. *IEEE Trans. on Commun.*, **COM-29**(9), 1290–1300.

Goeldner, E-H. (1985). An integrated circuit packet switching LAN: performance analysis and comparison of strategies. *CN and ISDN Systems*, **3–4**(10), 211–19.

Gonsalves, T.A. (1983). Packet-voice communication on an Ethernet local computer network: an experimental study. Paper presented to ACM SIGCOMM'83, Austin, Texas, March 1983.

Green, L. (1980). A queuing system in which customers require a random number of servers. *Operational Res.*, **28**, 1335–46.

Green, L. (1981). Comparing operating characteristics of queues in which customers require a random number of servers. *Management Sci.*, **27**(1), 65–74.

Gross, D. and Harris, C.M. (1985). *Fundamentals of Queuing Theory*. John Wiley, New York.

Harita, B.R. (1991). Dynamic bandwidth management. Technical Report no. 217, University of Cambridge, Computer Laboratories.

Harita, B.R. and Leslie, I.M. (1989). Dynamic bandwidth management of primary rate ISDN to support ATM access. Paper presented to SIGCOMM'89, Austin, Texas, 19–22 September 1989.

Henriet, D. (1987). PBXs and LANs in ISDN context. Paper presented to ISDN'87, 16–18 June 1987 (Online Computer Communications Series 12), London.

Heyman, D.P. (1968). Optimal operating policies for M/G/1 queuing systems. *OR*, **16**(2), 362–82.

Hilal, W. and Liu, M.T. (1984). Local area networks supporting speech traffic. *Computer Networks*, **8**(4), 325–37.

Ibe, O.C. and Gibson, D.T. (1986). Synchronous communication protocols for LANs. *CN and ISDN Systems*, **11**(2), 89–97.

IEEE (1988a). IVD LAN Interface Working Group: LAN/ISDN interconnect via frame relay, IEEE-802-9, Zurich (March).

IEEE (1988b). Part D, MAC Bridges, Revision E, IEEE-802.1A (D5-88/36) (May).

IEEE (1988c). Working Group, proposed IEEE Standard 802.6: Distributed Queue Dual Bus (DQDB)—Metropolitan Area Network (MAN), Draft D.0 (July), IEEE-802.6.

Ishii, H. (1988). ISDN user–network interface management protocol and its application. Paper presented to Globecom'88, Hollywood, Florida, 28 November–1 December 1988.

ISO (1984). Basic reference model for OSI. ISO 7498.

ISO (1986). Information processing systems: open systems interconnection—transport protocol specification. ISO 8073.

ISO (1987a). Network service definition. ISO 8348.

ISO (1987b). Network service definition—Addendum: CL-mode. ISO 8348 AD1.

ISO (1987c). Internal organisation of the network layer. ISO 8648.

ISO (1987d). Use of X.25 to provide the OSI connection-mode NS. ISO 8878 (1987-09-01).

ISO (1988). Protocol for providing the connectionless-mode network service. ISO 8473, (1988-12-15).

ISO (1989). Information processing systems:- data communications—network/transport protocol interworking specification. PDTR 10172 (November).

Kashper, A.N. (1987). Bandwidth allocation and network dimensioning for international multiservice networks. Paper presented to the ITC-5, Lake Como, Italy, 4–8 May 1987.

King, P.J.B. (1990). *Computer and Communication Systems Performance Modelling*. Prentice-Hall, Hemel Hempstead.

King, P.J.B. and Shacham, N. (1986). Queuing models for buffers with dial-up servers. Paper presented to IEEE INFOCOM'86, Miami 8–10 April 1986.

King, P.J.B. and Shacham, N. (1989). Queuing analysis for buffers with dial-up servers. *Perf. Eval.*, **10**(2), 129–45.

Kirstein, P.T., Deniz, D.Z., Knight, G.J. and Wilbur, S.R. (1989). The potential impact of the ISDN on JANET over the period 1989–1993 in the context of international open system standards. UCL Technical Report TR 159 (February).

Kleinrock, L. (1974). Resource allocation in computer systems and computer-communication networks. *Proc. of IFIP'74*, North-Holland, Amsterdam, pp. 11–18.

Kleinrock, L. (1976). *Queuing Systems*, Vol. II: *Computer Applications*, John Wiley, New York.

Knight, G.J. (1989). Project PROOF: primary rate ISDN in the office. Paper presented to the Online ISDN'89 Conference, London, June 1989.

Knight, G.J. and Kirstein, P.T. (1987). Project INCA: an OSI approach to office communications. Paper presented to the ISDN'87, London, June 1987.

Knight, G., Deniz, D. and Fan, J. (1987). Gateways to the ISDN. Paper presented to the ISDN'87, London, 16–18 June 1987.

Kobayashi, H. (1978). *Modeling and Analysis*, Addison-Wesley, Reading, Mass.

Kraimeche, B. (1984). Traffic access control strategies in integrated service digital networks. Ph.D dissertation, Columbia University.

Kraimeche, B. and Schwartz, M. (1984). Circuit access control strategies in integrated digital networks. Paper presented to the IEEE Infocom'84, San Diego, April 1984.

Kraimeche, B. and Schwartz, M. (1985a). Queue service policies for access control in ISDN. Paper presented at the ICC'85, Chicago, June 1985.

Kraimeche, B. and Schwartz, M. (1985b). Analysis of traffic access control strategies in integrated service networks. *IEEE Trans. Commun.*, **COM-33**(10), 1085–93.

Kraimeche, B. and Schwartz, M. (1986). Bandwidth allocation strategies in wide-band integrated networks. *IEEE JSAC*, **SAC-4**(6), 1327–35.

Kraimeche, B. and Schwartz, M. (1987a). A channel access structure for wideband ISDN. Paper presented at the IEEE Infocom'87, San Francisco, April 1987.

Kraimeche, B. and Schwartz, M. (1987b). A channel access structure for wideband ISDN. *IEEE JSAC*, **SAC-5**(8), 1327–35.

Kummerle, K. (1974). Multiplexer performance for integrated line- and packet-switched traffic. ICCC'74, Stockholm, August 1974.

Lai, W.S. (1989a). Frame Relaying Service: An Overview. Paper presented to the IEEE INFOCOM'89, Ottowa, 23–27 April 1989.

Lai, W.S. (1989b). Frame relaying: an ISDN additional packet mode bearer service. Paper presented to the IEEE ICC'89 Conference, Boston, Mass., 11–14 June 1989.

Lai, W.S. (1989c). Network and nodal architectures for the interworking between the frame relaying services. *ACM CCR*, **19**(2), 72–84.

Lamont, J., Doak, J. and Hui, M. (1989). LAN interconnection via frame relaying. Paper presented to the IEEE INFOCOM'89, Ottawa, 23–27 April 1989.

Lavenberg, S.S. (ed.) (1983). *Computer Performance Modeling Handbook*, Academic Press, London.

Leiner, B.M., Cole, R., Postel, J. and Mills, D. (1985). The DARPA internet protocol suite. *IEEE Commun. Mag.*, **23**(3), 29–34.

Lenzini, L. (ed.) (1984). Network interconnection: principles, architecture and protocol implications. Cost-11 bis LAN Group, Final Report (Part 1). CNR-Institute CNUCE.

Leslie, I.M., Needham, R.M., Adams, G.C. and Burren, J.W. (1984). The architecture of the universe network. Paper presented to the ACM SIGCOMM, Montreal, June 1984.

Levy, Y. and Yechiali, U. (1975). Utilization of idle time in an M/G/1 queuing system. *Man. Sci.*, **22**(2), 202–11.

Li, S.Q. (1988). Overload control in a finite message storage buffer. Paper presented to the IEEE INFOCOM'88, New Orleans, 29–31 March 1988.

Li, S. and Mark, J.W. (1984). Performance of integrated services on a single TDM system. ICC'84 Conference.

Liao, K.-Q., Dziong, Z. and Mason, L.G. (1989). Dynamic link bandwidth allocation in an integrated services network. Paper presented to the IEEE ICC'89 Conference, Boston, Mass., 11–14 June 1989.

Linington, P.F. (1983). Fundamentals of the layer service definitions and protocol specification. *Proc. of IEEE*, **71**(12), 1341–5.

Logistics Management Institute (LMI) (1989). CALS Telecommunications Plan (19 September).

Logistics Management Institute (LMI) (1990). CALS Telecommunications Planning Concepts (21 September).

Lu, F.V. and Serfozo, R.F. (1984). M/M/1 queuing decision processes with monotone hysteretic optimal policies. *OR*, **32**(5), 1116–32.

MacDougall, M.H. (1987). *Simulating Computer Systems: Techniques and Tools*. MIT Press, Cambridge, Mass.

Maglaris, B. and Schwartz, M. (1979). Optimal bandwidth allocation in integrated line- and packet-switched channels. Paper presented to the IEEE ICC'79 Conference, Boston, Mass.,10–14 June.

Maglaris, B. and Schwartz, M. (1982). Optimal fixed frame multiplexing in integrated line- and packet-switched communications networks. *IEEE Trans. on Information Theory*, **IT-28**(2), 263–73.

Meijer, A. (1987). *Systems Network Architecture*. Pitman, London.

Minzer, S.E. (1989). Broadband ISDN and asynchronous transfer mode (ATM). *IEEE Communications*, **29**(9), 17–24/57.

Mitra, N. and Usiskin, S.D. (1991). Relationship of the signalling system no. 7 protocol architecture to the OSI reference model. IEEE Network, 26–37.

Mitrani, I. (1982). *Simulation Techniques for Discrete Event Systems*. Cambridge University Press.

Moder, J.J. and Phillips Jr, C.R. (1962). Queuing with fixed and variable channels. *Opns. Res.*, **10**, 218–31.

Mood, A.M., Graybill, F.A. and Boes, D.C. (1974). *Introduction to the Theory of Statistics*, McGraw-Hill, New York.

Nelson, R.D. and Philips, T.K. (1989). An approximation to the response time for shortest queue routing. *ACM Performance Evaluation Review*, **17**(1), 181–9.

Nutt, G.J. and Bayer, D.L. (1982). Performance of CSMA/CD networks under combined boice and data loads. *IEEE Trans. on Commun.*, **COM-30**, 6–11.

Neuts, M.F. (1971). A queue subject to extraneous phase changes. *Adv. Appl. Prob.*, **3**, 78–119.

O'Prey, K. (1986). Putting factories on the MAP. *Computer Systems*, **6**, 31–4.

Peel, E.J.E. (1989). International extension of ISDN and its CPE implications. *Telecommunications*, **23(7)**, 61–5.

Perlman, R., Harvey, A. and Varghese, G. (1988). Choosing the appropriate ISO layer for LAN interconnection. *IEEE Network*, **2**(1), 81–6.

Postel, J.B. (1980). Internetwork protocol approaches. *IEEE Trans. on Commun.*, **28**(4), 604–11.

Postel, J. (1981a). Internet protocol: request for comment (RFC 791) (September).

Postel, J. (1981b). Transmission control protocol: request for comment (RFC 793) (September).

Potter, R.M. (1985). ISDN protocol and architecture models. *CN and ISDN Systems*, **10**(3-4), 157–86.

Prue, W. and Postel, J. (1988). A queuing algorithm to provide type of service for IP links: request for comment (RFC 1046) (February).

Quarterman, J.S. and Hoskins, J.C. (1986). Notable computer networks. *Commun. of the ACM*, **29**(10), 932–71.

Roberts, J.W. and Liang, Y. (1988). Performance of the multilink protocol for Nx64 kbps data transmission in ISDN. Paper presented at the International Seminar on Teletraffic, Beijing, September 1988.

Roberts, J.W. and Van, A. Hoang (1987). Characteristics of services requiring multi-slot connections and their impact on ISDN design. Paper presented at the ITC Seminar, Lake Como, May 1987.

Romani, J. (1957). Un Modelo de la teoria de colas con numero variable de canales. *Trabajos Estadistica*, **8**, 175–89.

Ross, F.E., Hamstra, J.R. and Fink, R.L. (1990). FDDI—A LAN among MANS. *ACM SIGCOMM CCR*, **20**(3), 16–31.

Saaty, T.L. (1961). *Elements of Queuing Theory*, McGraw-Hill, New York.

Sastry, A.R.K. (1984). Performance objectives for ISDNs. *IEEE Communications Magazine*, **22**(1), 49–55.

Schlanger, G.G. (1986). An overview of signalling system no. 7. *IEEE JSAC*, **4**(3), 360–5.

Schneidewind, N.F. (1983). Interconnecting local networks to long-distance networks. *Computer*, **16**(9), 15–24.

Schwartz, M. (1987). *Telecommunications Networks; Protocols, Modeling and Analysis*. Addison-Wesley, Reading, Mass.

Schwartz, M. and Kraimeche, B. (1982). Comparison of channel assignment techniques for hybrid switching. Paper presented at the IEEE, Philadelphia, June 1982.

Seifert, W.M. (1988). Bridges and routers. *IEEE Network*, **2**(1), 57–64.

Serres, Y. De and Mason, L.G. (1988). A multiserver queue with narrow- and wide-band customers and wide-band restricted access. *IEEE Trans. on Commun.*, **36**(6), 675–84.

Sorensen, S-A. (1988). The SLOANE network emulator. UCL Indra Note.

Sunshine, C.A. (1990). Network interconnection and gateways. *IEEE JSAC*, **8**(1), 4–11.

Tanenbaum, A.S. (1989). *Computer Networks*, 2nd edn. Prentice-Hall, Englewood Cliffs, NJ.

Tashman, L.J. and Lamborn, K.R. (1979). *The Ways and Means of Statistics*. Harcourt Brace Jovanovich, New York.

Tennenhouse, D.L. and Leslie, I.M. (1989). A testbed for wide area ATM research. Paper presented at the SIGCOMM'89, Austin, Texas, 19–22 September 1989.

Tennenhouse, D.L., Leslie, I.M., Needham, R.M., Burren, J.W., Adams, C.J. and Cooper, C.S. (1987). Exploiting wideband ISDN: the Unison exchange. Paper presented at the IEEE INFOCOM'87 Conference, San Francisco, 31 March–2 April 1987.

Tobagi, F.A. and Cawley, N.G. (1982). On CSMA/CD local networks and voice communications. Paper presented at the IEEE INFOCOM'82 Conference, Las Vegas, 30 March–1 April.

Tran-Gia, P. and Rathgeb, E. (1987). Performance analysis of semidynamic scheduling strategies in discrete-time domain. Paper presented at the IEEE INFOCOM'87 Conference, San Franciso, 31 March–2 April 1987.

Trudgett, P.A. (1988). ISDN: its impact on OSI data communications. *Proceedings of International Open Systems*, Online, London, pp. 389–401.

Walton, S.M., Knight, G.J., Carbonell, R. and Hardie, M. (1991). The construction of a connectionless Ethernet ISDN gateway. *Microprocessors and microsystems*, **15**(1), 3–10.

Weinstein, C.J., Malpass, M.L. and Fisher, M.J. (1980). Data traffic performance of an integrated circuit- and packet-switched multiplex structure. *IEEE Trans. Commun.*, **COM-28**(6), 873–8.

Wilford, B.A. (1984a). CMOS user guide. UCL Internal Note 1660.

Wilford, B.A. (1984b). CMOS system calls. UCL Internal Note 1661.

Wong, J.W. and Gopal, P.M. (1984). Analysis of a token ring protocol for voice transmission. *Computer networks*, **8**(4), 339–46.

Yadin, M. and Naor, P. (1963). Queuing systems with a removable service station. *ORQ*, **14**(4), 393–405.

Yadin, M. and Naor, P. (1967). On queuing systems with variable service capacities. *NRLQ*, **14**(1), 43–54.

Yechiali, U. and P. Naor (1971). Queuing problems with heterogeneous arrivals and service. *OR*, **19**(3), 722–34.

Yum, T. and Schwartz, M. (1987). Packet-switched performance with different circuit-switched routing procedures in nonhierarchical integrated circuit-switched and packet-switched networks. *IEEE Trans. on Commun.*, **COM-35**(3).

Zacks, S. and Yadin, M. (1970). Analytic characterization of the optimal control of a queuing system. *J. Appl. Prob.*, **7**, 617–33.

Zafiropulo, P. (1974). Flexible multiplexing for networks supporting line-switched and packet-switched data traffic. Paper presented at the ICCC'74, Stockholm, August 1974.

Zukerman, M. and Rubin, I. (1986). On multi-channel queuing systems with fluctuating parameters. Paper presented at the IEEE INFOCOM'86, Miami, 8–10 April 1986.

Zylberberg, T. (1986). PABXs, LANs, ISDN: how do they fit together? *Proc. Int. Conf. on ISDN (ISDN'86)*, **1**, 109–20 (Online Computer Communications Series : 4).

GLOSSARY

Address resolution protocol (ARP)
The internet protocol which is used to map internet addresses to physical addresses on local area networks in a dynamic way. The network must support hardware broadcast feature.

A priori
With previous knowledge, or a previous agreed set of procedures.

Application program interface (API)
A software interface which provides a standard method of usage for a particular software package or module.

Asynchronous transfer mode (ATM)
A data transportation mode in which messages are placed into fixed sized cells. In the asynchronous mode, the message sources are not allocated fixed time slots, as in the case of time division multiplexing, for their transfers: rather, they occupy the bandwidth only whenever they send cells.

Bandwidth
The difference between the highest and lowest frequencies of a continuous frequency spectrum. It is also used to indicate capacity of a communications channel.

Bandwidth allocation
Assignment of basic bandwidth units to different sources of traffic.

Bandwidth management
Management of the bandwidth resource at the access point to a communications facility.

Basic bandwidth unit (BBU)
Smallest unit of allocatable bandwidth.

Basic-rate access (BRA)
A standardized access configuration in ISDNs with an information-carrying capacity of 144 kbps organized in two B+D-channel groupings. Each B-channel carries user information at 64 kbps while the D-channel is a 16 kbps channel used for signalling and low-speed data. With framing, synchronization and echo bits, the aggregate transmission rate of the BRA is 192 kbps.

Basic-rate interface (BRI)
The physical and logical access point to the basic rate access in ISDNs.

Bearer service
A telecommunications facility providing exchange of information between access points to a network.

Bit error rate (BER)
The percentage ratio of the bits received with error to the total bits received.

Bit pipe
A term used to refer to a circuit-switched path with unrestricted information transfer and 8 kHz structural integrity capabilities over which transfer of user information is achieved without alteration. The network provides only the physical layer, and the user is responsible for the provision of higher-layer protocol functions.

Broadband ISDN (B-ISDN)
The second-generation ISDN providing wideband capabilities and supporting channels with bit rates greater than that of the primary rate.

Bridge
A device used for connecting two or more physical networks at the physical and data link layers. It forwards the packets that arrive from the connected networks to their destinations. It is also referred to as the data link layer relay or intermediate system in the OSI terminology.

Burst
A short spurt of data (packets).

Bursty
A traffic source or stream that has bursts.

Burstiness
The measure providing an idea about how bursty a traffic source is.

Cell relaying
See Fast Packet Switching.

Channel (Transmission channel)
A means of unidirectional transmission of signals between two points. Several channels may share a common path; e.g. each channel may be allocated a particular frequency band or a particular time slot.

Channel assignment
The process of assigning communication requests to the available channels at the access point to a transmission facility.

Channel management
The terminology used to indicate the effective management of the multiple physical and or logical channels at the access point to a transmission facility.

Channel status table (CST)
A table containing the current status of all the individually available physical and or logical channels at the user–network interface.

Circuit
A combination of two transmission channels permitting bi-directional transmission of signals between two points, to support a single communication.

Circuit switching (CS)
A communications mode whereby a 'physical' circuit is established between the communicating parties prior to the transmission of information. The circuit is held for the duration of the communications session.

Common channel signalling (CCS)
A technique whereby the signalling information relating to a multiplicity of communicating entities is transmitted separately from the user information channels on a special signalling channel using pre-defined and addressable messages.

Connection
A physical or logical association between two communicating entities.

Connectionless (CL) mode
A communications mode where the communicating entities do not need to set up a call on an end-to-end basis.

Connection oriented (CO) mode
A communications mode where the communicating entities need to set up a network connection before being able to exchange information. The connection is released at the end of the communication.

Customer
In queuing theory, this is the next arrival at the queuing system requesting a service. In data communications terms, it is a protocol data unit (PDU), packet frame or bit stream arriving at a transmission facility.

Customer premises equipment
Telecommunications equipment located at the user premises but technically belonging to the service provider. It may include network termination, PABXs and wiring.

Data circuit terminating equipment (DCE)
A communications equipment that interfaces the user equipment (DTE) and the transmission line by providing the necessary coding and signal conversion between them.

Data link layer (DLL)
The second layer of the ISO OSI reference model. It provides error handling and synchronization, making a reliable transmission possible on unreliable communication channels.

Data terminal equipment (DTE)
A terminating equipment situated at the end of a communications path where the information is converted from the user-provided form to the transmission signals and vice versa. It is the counterpart of a data circuit-terminating equipment.

Device driver
This is a software module designed to interface and facilitate the control of devices connected to a host by interfacing with the operating system on one side and the physical device on the other. Device-specific features are made transparent to the operating system by providing a standard user or application program interface (API).

Dynamic bandwidth management (DBM)
Managing the bandwidth available at the access point to a communications facility adaptively as the traffic conditions change in time.

Dynamic channel management (DCM)
Managing the multiple channels available at a user–network access point adaptively within a dynamic environment.

Encapsulation
A process whereby the packets belonging to one protocol are transported across a network operating a different protocol by placing them as payload within the packets of the crossed network. This avoids protocol conversion at the network interconnection point. Also known as **tunnelling** (see below).

End system (ES)
A host or device attached to an OSI network acting as the terminating system for OSI protocols at each of the seven layers.

Entity
A unit within a layer which performs the layer functions within an end system, accessing the layer entities below and providing services to the layer entity above at the service access points.

Fast packet switching (cell relaying)
A new wide area networking technology based on the idea of switching and transmitting fixed-length short packets (or cells) at high speed. It is capable of transmitting digital data, digitized voice and digitized video. Fast switching speed is achieved by operating the switches at layer 2 of the OSI model and hence avoiding overheads associated with layer 3.

Fast select
An optional feature in the X.25 protocol that allows the inclusion of user data in the call set-up and clearing packets.

Frame
A group of de-limited bits transmitted over a link. It may contain control, synchronization, addressing and error detection/correction information and its size is dependent on the protocol used.

Frame relaying
A technique whereby the frames are sent along a transparent network path established at the call set-up phase after checking only the frame format. Frame order is preserved and no acknowledgements, are used. For detection, throughput and congestion control are to be studied.

Frame switching
A technique similar to frame relaying except that both the user and the network must implement the full layer 2 protocol providing frame acknowledgement, flow control, error detection and recovery as well as frame format check.

Gateway
A generic name for referring to the data link and above layer relays. More appropriately, it is a network interconnection device operating at layer 7 of the OSI model. In the OSI terminology, it is an intermediary system or interworking unit (IWU).

High-level data link control (HDLC)
A layer 2 (data link layer) bit-oriented protocol providing the data link service to the network layer.

Host
A computing system attached to a network and providing services to the network users.

Hysteresis
A relationship between two variables in which the increase and decrease in the value of one variable causes two different and nonlinear variations in the value of the second.

Integrated digital network (IDN)
A network using digital technology and integrating the switching and transmission functions, thereby achieving the provision of better services in a circuit-switched telecommunications network.

Integrated services digital network (ISDN)
An evolving global telecommunications network that will provide end-to-end digital connectivity and multi-service capability (voice, data and video) through standard user interfaces.

Integrated services local network (ISLN)
A local network technology that can handle multiple services like voice, data and video within the same network.

Integrated services PBX (ISPBX)
A private branch exchange having a multi-service capability. Most ISPBXs support ISDN standards and protocols.

Interconnection (Network interconnection)
A term usually applying to the network concatenation such that hosts and users on different networks can communicate through intermediate systems like relays (gateways).

Intermediate system (IS)
An OSI terminology used to indicate a device or a system which acts as an intermediary in network interconnection as opposed to an end-system. It does not terminate OSI protocols but acts as a relay. *See* **Router**, **Bridge** and **Repeater**.

Internet protocol (IP)
The network layer (layer 3) protocol for the internet protocol suite. It is based on the connectionless mode and uses datagrams.

Internetworking
Concatenation of (computer) networks using inter-working units (IWUs) which have access to two or more networks and provide routing and relaying between them.

Interworking
Provision of services and facilities available on one network (or communications system) to the users on another network or to users on the same network using isolated terminals or workstations.

Local area network (LAN)
A communications network connecting computer equipment within a limited geographical area and achieving resource-sharing and user communications.

Logical channel
A channel of arbitrary bandwidth composed of one or more (physical) channels.

Medium access control (MAC)
In a multi-device network, the method of determining which device has access to the transmission medium. For the IEEE LANs, this function is implemented in the MAC sublayer of the data link layer (layer 2) of the OSI model.

Metropolitan area network (MAN)
A communications network designed to link devices within a metropolitan or campus geographical area and offering high-speed access and transmission, e.g. the IEEE 802.6 MAN standard, based on the distributed queue dual bus (DQDB) technology.

Multi-link protocol (MLP)
A feature of the CCITT X.25 standard which allows the setting up of multiple data links across single or multiple physical paths.

Narrowband ISDN (N-ISDN)
The first generation ISDN which can support data transmission rates at up to 2.048 Mbps.

Network interconnection
See **Interconnection.**

Network layer (NL)
Layer 3 of the OSI Reference Model. It is responsible for connection set-up, maintenance and termination between end systems. It also supports addressing and routing functions.

Network terminating equipment (NTE)
A functional grouping at the ISDN user–network interface (UNI) associated with termination (network termination type 1), distribution and switching (network termination type 2) functions.

Open systems interconnection (OSI) Reference Model
A conceptual model for communications between co-operating devices, where the communications functions are placed in a seven-layer architecture.

Packet (data)
A group of bits which are de-limited and may include data, address and/or control information. It usually has header, payload and trailer sections. A packet is switched, routed, relayed and transmitted as a unit of information.

Packet level protocol (PLP)
Layer 3 protocol of the CCITT X.25 standard.

Packet switching
A method of information exchange where long messages are subdivided into packets and are then switched using the address information in the packet header. Transmitted messages travel between switching centres, where they may be stored very briefly and then routed on to the next transmission path until they reach destination.

Physical channel
See **Channel**.

Physical layer
Layer 1 of the OSI model, responsible for the mechanical and electrical interfacing to the communications medium. It converts data bits into appropriate signals capable of being transmitted over the communications medium and vice versa.

Point-to-point
A configuration in communications where a transmission path is shared between the two communicating devices or entities.

Porting
Applying, using, transferring and, if necessary, adapting an application of software from one environment to another.

Primary-rate access (PRA)
A standardized access configuration in ISDNs with an information-carrying capacity of 2.048 Mbps in Europe or 1.544 Mbps in the United States and Japan. It can be organized into 30 B+D (in Europe) or 23 B+D (in United States and Japan) channel groupings. Other channel groupings are also possible.

Primary-rate interface (PRI)
A user–network interface configuration in ISDN providing primary-rate access.

Protocol
A formal description of the set of rules that provide ordered communications between layer functions.

Protocol data unit (PDU)
Information exchanged between peer entities via a common protocol. It may carry user data, control or address information.

Quality of service (QoS)
A collection of properties associated with the service obtained from a communications facility. The attributes of QoS are different for different modes of operation and vary from network to network.

Reference configuration
A general configuration of network functions grouped together and separated by reference points.

Reference point
An imaginary point at the place of attachment between functional groupings which is used to make references to the interface thus formed.

Relay
An intermediate system (IS) in the OSI terminology, used for interfacing two or more subnetworks.

Repeater
A device that operates at layer 1 of the OSI model and extends the maximum length of cable that can be used in a single network.

Router
A network layer (layer 3) intermediate system (or network layer relay) in the OSI terminology. It is used for network interconnection. A router, routes packets to its appropriate network interface for transmission on the next hop of their journey.

Routing table
A table that is kept and managed in a router device and consulted when packets are routed. It contains entries of network addresses, interface (port) identification, some routing metric and some other implementation specific entries.

Scheduling
Refers to packet scheduling in a data communications facility. It is the policy used in assigning a packet to one of several transmission queues (lines) on its arrival. Examples of scheduling policies are: round robin, random and shortest queue, etc.

Scheduling policy
See **Scheduling**.

Server
This is the unit of a queuing system which deals with each arriving or queuing customer in accordance with a service policy (e.g. first come–first served). In packet data devices this is the hardware and software element responsible for transmitting packets on the communications path.

Signalling protocol
The protocol used on a signalling channel. It usually refers to outband (and common channel) signalling systems.

Status table
See **Channel status table**.

Store and forward
A technique used in packet data communications in which arriving packets are stored in the buffers of a switching device and then forwarded onward towards their destination. It is used in both the message and packet-switched networks.

Stream
(with reference to traffic) Steady flow.

Subnetwork
A group of OSI end and intermediate systems controlled from a single administrative domain. The network access protocol used is common to all users.

Superchannel
A channel made up by the aggregation of basic channel types (e.g. in ISDN primary-rate interface; nB + mH-channels) and having a bandwidth that is an integral multiple of the basic bandwidth unit (e.g. $n \times 64$ kbps).

Switching
The process of interconnecting functional units, transmission channels or circuits for the duration of a communications session.

Synchronous transfer mode (STM)
A transfer mode which offers transmission capacity of a fixed-length word periodically to each user, process or entity.

Time division multiplexing (TDM)
A method of sharing the bandwidth of a communications facility between two or more sources by allocating the transmission facility to different channels for a duration of time.

TDM frame
A unit of time within which all the TDM time slots occur and are repeated in consecutive time slices.

Telecommunication
Conveyance of information encoded into signals using electric, optic or electro-magnetic media.

Time slot (TS)
A unit of time within a time division multiplex which is repeated in consecutive frames. The association of the same or different time slots in consecutive (synchronous) frames gives rise to channels.

Time slot interchange
Interchanging of time slots within a time division multiplex frame.

Time slot sequence integrity (TSSI)
Preservation of the ordered sequence of time slots from source to destination. The time slot sequence integrity can be lost as a result of the time slot interchange at intermediate switches and delays sustained owing to transmission on different paths.

Transmission
Conveyance of information encoded into signals between two or more points.

Transmission channel
See **Channel**.

Transmission control
The control involved in the act of de-queuing a data packet or frame from a buffer and transmitting it across a communications path.

Transmission control protocol (TCP)
A transport layer (layer 4) protocol belonging to the internet suite of protocols and providing reliable, connection oriented and full duplex end-to-end communications.

Tunnelling
A method used in data communications where the packets belonging to one protocol and operating 'end-to-end' are transparently carried across a network or between communicating entities not supporting the protocol of the carried payload—for example 'IP tunnelling' of X.25.

User
A person, machine or process using the facilities and services of a communications or computer network or device.

User–network interface (UNI)
An interface at the connection point between the user equipment and the network terminating equipment. User data and access protocols are transferred across it.

Value added services (VAS)
Services built upon the basic services provided by a telecommunications network and usually sold to the public by value added network operators.

Virtual circuit (VC)
A packet-switching technique whereby a connection is established between the two end parties before data can be exchanged. It provides the advantages of a point-to-point link such that, after the establishment of the path, only a virtual circuit identifier is needed rather than retention of the complete address of the destination and message sequence is inherently provided.

INDEX